Space: The Fragile Frontier

Space: The Fragile Frontier

Mark Williamson
BSc, CPhys MInstP, CEng MIEE, FBIS
Space Technology Consultant

Published by
American Institute of Aeronautics and Astronautics, Inc.,
1801 Alexander Bell Drive, Reston, VA 20191-4344

American Institute of Aeronautics and Astronautics, Inc., Reston, Virginia

1 2 3 4 5

Library of Congress Cataloging-in-Publication Data on file.
ISBN 1-56347-776-9

Front Cover:
Shadows cast by Saturn's rings on the planet's cloud tops form a backdrop to the moon Mimas in this stunning true-colour view from NASA's Cassini spacecraft. Mimas is 400 km across and orbits 185,000 km above the planet. Cassini was 1.4 million km from Saturn when it took this image on 18 January 2005. [NASA/JPL/Space Science Institute]

Back Cover:
Upper image: The ultimate iconic image of the 20th century – but how long will Aldrin's boot print survive once the lunar tourists arrive? [NASA]
Lower image: A 'frozen sea' near the Martian equator – part of the Martian environment worth protecting. [ESA/DLR/FU Berlin (G. Neukum)]

To Rita, Matthew, and Thomas.
Three stars in a fragile universe.

ACKNOWLEDGMENTS

I am indebted to a number of friends and colleagues for the time they gave so freely in reviewing sections of the original manuscript and would like to acknowledge their contributions, which have improved the book. They are not, of course, accountable for the text as it appears, and any errors are entirely my own responsibility.

I would particularly like to thank Nicholas Johnson, who, in addition to making useful comments and corrections, supplied a number of graphical illustrations and data for the tables. I am also grateful to Ivan Almar, Frances Brown, Walter Flury, Mark Lupisella, Frank Lyall, Wendell Mendell, Bess Reijnen, Rick Steiner, Patricia Sterns, Les Tennen, and David Wade, whose contributions were invaluable.

For their help in illustrating the book, I would also like to thank Fernand Alby, Charles Barder, David Hardy, and Alistair Scott for specific permission to use images. I would also like to acknowledge the GRIN (Great Images in NASA) database and the ESA picture archive.

Finally, I would like to thank James Van Allen for sparing the time to write a foreword to this volume, and of course my wife, Rita, for her usual dedication in support of my writing projects.

TABLE OF CONTENTS

PREFACE

I first became interested in protection of the space environment at the 1989 IAF Congress in Malaga, hopping between sessions on space debris and space law. One notable paper on the chain reaction or "cascade effect," whereby debris objects hit other objects to produce yet more debris, raised the possibility that some of the more useful orbits could become inaccessible within a few decades. Meanwhile, the technicalities of debris production and its ramifications for commercial operations were being voiced—somewhat tentatively—to an audience of space lawyers. It was clear, however, that there was little, if any, interaction between engineering and law communities.

Outside of those sessions, among the mainstream conference topics, there appeared to be little recognition of the existence of a problem, let alone the need for a solution. Despite the fact that the damage caused to orbiting spacecraft by debris from other spacecraft had already been recorded and analyzed, the space community at large seemed content to ignore it.

At the following year's congress, in Dresden, I made my first attempt to engage with space lawyers by offering a logical framework for definitions of the most common terms in space technology—a basic structure for a common language.[1] This was partly in response to a call from some lawyers for a greater contribution from the space technology fraternity, but was also intended as a modest attempt to help them place their discussions in the proper technical context.

It was evident that parts of existing space law, having been written in the late 1960s, failed to address modern-day—let alone future—requirements, at least in part because of a lack of technical input. It was also clear that laws or policies relating to orbital debris were only the beginning of the story. Pollution of the Martian environment was already of concern to some of the more forward-looking thinkers (among both science/technology and law communities). Indeed, I should say that many of the people quoted in this book have been writing about the subject much longer than I have, and I am indebted to them for opening my eyes to some of the issues involved.

However, it was only in 1996, while researching a paper on the history of space science missions,[2] that I realized how much hardware had been crashed onto the Moon in the name of space research. The impression that planetary bodies beyond the Earth were not valued for their inherent "geo-morphological qualities" was augmented by the announcement that Japan's Lunar-A spacecraft would carry surface penetrators, which would add to the catalog of man-made lunar debris.

This led me to prepare a paper for the International Institute of Space Law on protection of the space environment—or lack of it—under the Outer Space Treaty.[3] It was accompanied by the publication of a number of maga-zine editorials and commentaries that strove to publicize my concerns to a wider audience of space professionals. Some of this appeared under the intentionally provocative heading "Are We Trashing the Solar System?,"[4-8] specifically with the intention of "sounding out" the respective readerships. I found the reaction both interesting and informative. Moreover, the responses served to convince me that a program to increase awareness of the issue was justified and necessary.

For example, a respondent to the piece in *Space News* was "saddened" by it, finding my concerns about lunar spacecraft debris "laughable."[9] In response to the question of whether the space environment constitutes "a body of resources to be plundered," he wrote, "the answer is: of course," adding that it was "hopeless to attempt to regulate it." He concluded that "it might just be worth emigrating to Mars just to get away from meddlesome environmentalists who share Mr. Williamson's beliefs. Let them create an environmentalist utopia on Earth; the rest of us will be busy 'exploiting' the solar system." A communication received in response to that critique was somewhat more heart-ening: the critique provided "the best argument I have seen to date for taking the stance you have put forward," wrote my correspondent, adding: "I fear his mentality is typical but I hope not endemic within the space industry."

A respondent to the piece in *Spaceflight* magazine disagreed that the lunar environment was worth protecting in its current state, suggesting—somewhat perversely I felt—that it should only be protected "after several thousand years of intensive industrial activity."[10] On the other hand, another respondent disagreed with my hesitation in recommending a legisla-tive solution, while opening my eyes to an archaeological theme. He would "criminalize the vandalism or theft of old spacecraft or their component parts with internationally binding legislation that reflects the statutory protection given to artefacts on terrestrial archaeological sites."[11] The debate I had evoked was illuminating, but it dried up almost as quickly as it had begun; it seemed that no one else was interested.

However, I soon realized that a Hungarian colleague, with whom I had been working on space terminology issues under the auspices of the Inter-national Academy of Astronautics (IAA) was also interested in protection of the space environment. Indeed, he had been for some time and had already suggested that the IAA form an international consultative study group. I asked him whether he thought there was any possibility of an

IAA symposium on "Protection of the Space Environment," and he suggested we propose an IAA/IISL Scientific-Legal Round Table on the subject, which was held at the 1999 IAF Congress in Amsterdam.

The objectives of the Round Table were to present the facts regarding actual and potential degradation of the space environment by human activity, particularly that of the planetary bodies; to allow discussion of the problems; and to consider how awareness of the problems might be enhanced among the space community. It attracted a good audience, and the rapporteur optimistically reported that the event could be regarded as "one first step to turn around the apparent lack of interest . . . displayed by the Member States of the UN." I have since come to realize that it was a very small step indeed.

This brings us, in a somewhat roundabout way, to this book. Following further discussions with colleagues and the publication of a few more papers on the subject,[12-14] I decided in early 2002 to pitch a book to a number of space-aware publishers. As a result, I signed a contract with the AIAA in March 2005. I mention this detail only as an excuse to credit Rodger Williams, of AIAA Publications Development, who had the foresight to recognize the potential importance of the subject.

In one sense this book is ahead of its time, in that many space professionals remain to be convinced about the need to protect the space environment, but in another it is just in time because further delay might mean that protection comes too late.

The protection of the space environment—for the study of and use by successive generations of explorers and developers—is an important concept that has yet to enter the collective consciousness of the space community, let alone the wider terrestrial community. One of my intentions, in writing this book, is to illustrate the relevance of the space environment and explain why we should consider protecting some of its unique properties and its most significant territories.

There will always be those on the one side who seek to ridicule and devalue well-meant suggestions for improvement, and those on the other who exaggerate and emotionalize their claims to ensure an audience, but there is a middle way. This book attempts to take that middle way by presenting the facts, explaining the premises, and extrapolating into an uncertain future in a logical manner.

The majority of the book takes a standard form and style—presenting facts, arguments, and possible solutions (all duly referenced)—but in recognition of the power of predictive fiction, it includes three short chapters that present opposing scenarios for the future of space exploration and development. They are written, in fictional style, from the perspective of the year 2057—the centenary of the Space Age—offering three different views of our future in space. I make no apology for including fictional accounts in an otherwise serious book. It is clear that there are limitations to how far we can extrapolate what we know now into the future. Borrowing the methods of a different genre releases us from the burden of absolute proof, while liberating our minds to consider the what-if scenarios we might otherwise ignore.

This is the first book to draw together the recognized issues of Earth orbital debris and planetary protection, set them in the context of space law and ethical policies, and encourage a balance between desirable expansion into space and protection of the space environment. In essence, it calls for a sustainable and environmentally friendly approach to space exploration and development.

I hope that younger members of the space community will adopt this text as part of their battle cry for space exploration and development and recognize it not as a hindrance to progress, but as a model for thoughtful and considered expansion from the home planet. Whereas more senior members of the community might find difficulty in adapting to environmentally friendly models, those who have yet to manage their first project or rise to the ranks of principal investigator have grown up on a planet already steeped in environmental concerns. As a result, sustainability is ingrained rather than imposed. My hope is that they are the generation that will extend this recognition to the space environment and design their technology and their missions with this firmly in mind.

If this book serves to add justification to their ideas for sustainable growth and development through the solar system, and power to their elbow in arguing for environmental concern, that would be pleasing. But if it simply lifts their eyes to the potential damage to the space environment that can occur in the name of science and commerce, it will have served its purpose, because knowledge, once learned, cannot be unlearned. Whatever they do, I would urge them to explore and develop the new frontier, but to leave the less attractive habits of their forebears behind.

Go with care, consideration, and a sense of wonder . . . we are but guardians of the fragile frontier.

Mark Williamson
Kirkby Thore, Cumbria
February 2006

REFERENCES

[1]Williamson, M., "Space Law and Space Technology—A Common Language," *Proceedings of 33rd Colloquium on the Law of Outer Space*, 6–12 Oct. 1990.

[2]Williamson, M., "The Early Development of Space Science Satellites and Planetary Probes," Inst. of Electrical Engineers, July 1996.

[3]Williamson, M., "Protection of the Space Environment Under the Outer Space Treaty," International Inst. of Space Law, Paper IISL-97-IISL.4.02, Oct. 1997.

[4]Williamson, M., "The Other Debris," Editorial, *SPACE and Communications*, Vol. 13, Nov./Dec. 1997, p. 2.

[5]Williamson, M., "Are We Trashing the Solar System?," Commentary, *Space News*, Vol. 8, 1–7 Dec. 1997, p. 15.

[6]Williamson, M., "Protecting the Space Environment—Are We Doing Enough?," Viewpoint, *Space Policy*, Feb. 1998, pp. 5–8.

[7]Williamson, M., "Protecting the Space Environment," Into Space, *Spaceflight*, Vol. 40, No. 4, April 1998, p. 135.

[8]Williamson, M., "Are We Trashing the Solar System?," Perspective, *Earth Space Review*, Vol. 6, No. 4, 1998, pp. 22–24.

[9]Glass, J. F., "Solar Trash," Letters, *Space News*, Vol. 9, 12–18 Jan. 1998, p. 14.

[10]Ashworth, S., "Protect or Promote," Correspondence, *Spaceflight*, Vol. 40, No. 7, July 1998, p. 270.

[11]Fewer, G., "Space Heritage Sites," Correspondence, *Spaceflight*, Vol. 40, No. 8, Aug. 1998, p. 286.

[12]Williamson, M., "Exploration and Protection—A Delicate Balance," International Academy of Astronautics, Paper IAA-01-IAA.13.1.03, Oct. 2001.

[13]Williamson, M., "Protection of the Space Environment: the First Small Steps," Committee on Space Research, Paper COSPAR02-A-01364/PPP1-0014-02, Oct. 2002; also *Advances in Space Research*, Vol. 34, 2004, pp. 2338–2343, URL: http://www.sciencedirect.com.

[14]Williamson, M., "Space Ethics and Protection of the Space Environment," IAC-02-IAA.8.1.03, Oct. 2002.

FOREWORD

This important book provides an authoritative dose of prudence to the flourishing fields of space exploration and space technology.

The author's approach is avuncular. On the one hand he is deeply impressed by the diverse progress in space during the nearly 50 years since Sputnik 1. On the other hand he adopts the role of a kindly uncle who admonishes a profligate young nephew to mend his ways or suffer fatal consequences.

Early chapters give an up-to-date sketch of the achievements in space, which are, perhaps, sufficient to earn the designation The Space Age. For the lay person the best known of such achievements have been human space-flight and images of distant astronomical objects. But, arguably, far more important achievements have been the development of global telecommunications, surveillance, civil and military reconnaissance and navigation, and spectacular advances in science. The latter include improved understanding of the Earth as a durable habitat for humans and other species of life; the many terrestrial influences of the sun's radiation; the properties of the planets, their satellites and rings, and those of comets, asteroids and the interplanetary medium.

Hundreds of automated, commandable satellites and far-ranging spacecraft have been responsible for these scientific advances and for the pervasive utilitarian applications of space technology. But what of the future?

Mark Williamson proceeds to summarize the adverse effects of orbiting space debris, a most undesirable side effect of the international activity in space. He emphasizes the prospect that irresponsible contributions to the population of space debris, if continued, will destroy the viability of the most desirable orbits around the Earth and thus negate the future of space in the service of mankind. He makes an urgent plea for broad international recognition of this threat and for immediate and effective countermeasures. Otherwise the great promise of The Space Age will grind to an ignominious halt.

He also addresses the problems of forward contamination of the Moon, the planets, and their satellites, and the reverse contamination of the Earth by returning spacecraft with or without intended samples of extraterrestrial material. The matter of forward contamination has been highlighted by Robert Park of the American Physical Society in the following remark: "If any form of life is eventually found on Mars, it will probably be familiar—for the simple reason that it had been delivered there from the Earth."

The four concluding chapters of this comprehensive book are obligatory reading for every person who has a responsible role in space operations and/or space policy.

James A. Van Allen
University of Iowa
September 2005

James Van Allen is a Regent Distinguished Professor at the University of Iowa, whose research on energetic particles in space began with the use of high performance rockets in 1946. His radiation instrument in the first U.S. satellite Explorer 1 in 1958 revealed the existence of the radiation belts of the Earth, later called the Van Allen radiation belts. He has served as principal scientific investigator on 23 subsequent space missions including the first such missions to Venus, Mars, Jupiter, Saturn and the outer heliosphere.

ABBREVIATIONS AND ACRONYMS

ABM	= antiballistic missile
ACTS	= advanced communications technology satellite
ASAT	= antisatellite
ASTP	= Apollo–Soyuz Test Project
ATV	= Automated Transfer Vehicle
AU	= astronomical unit
CASC	= China Aerospace Science and Technology Corporation
CETEX	= Committee on Contamination by Extraterrestrial Exploration
CHM	= common heritage of mankind
CNAA	= China National Aerospace Administration
CNES	= Centre National d'Etudes Spatiales
CO_2	= carbon dioxide
COMEST	= Commission on the Ethics of Scientific Knowledge and Technology
COPUOS	= Committee on the Peaceful Uses of Outer Space
COSPAR	= Committee on Space Research
COSTIND	= Commission of Science, Technology and Industry for National Defence
CSM	= Command and Service Module
DART	= Demonstration of Autonomous Rendezvous Technology
DoD	= Department of Defense
EELV	= Evolved Expendable Launch Vehicle
ERTS	= Earth Resources Technology Satellite
ESA	= European Space Agency
ESOC	= European Space Operations Centre
ESTEC	= European Space Technology Centre
EVA	= extravehicular activity
FAA	= Federal Aviation Administration
FAI	= Fédération Aéronautique Internationale
FCC	= Federal Communications Commission
FOBS	= Fractional Orbital Bombardment Satellite
GEO	= geostationary orbit

GMD	=	ground-based midcourse defense
GMT	=	Greenwich Mean Time
GPS	=	Global Positioning System
HAPS	=	Hydrazine Auxiliary Propulsion System
HEO	=	high Earth orbits/highly elliptical orbits
HEPA	=	High-Efficiency Particulate Arrestor
HGS	=	Hughes Global Services
HST	=	Hubble Space Telescope
IAA	=	International Academy of Astronautics
IADC	=	Inter-Agency Space Debris Coordination Committee
IAF	=	International Astronautical Federation
IAU	=	International Astronomical Union
ICBM	=	intercontinental ballistic missile
ICO	=	intermediate circular orbit
ICSU	=	International Council of Scientific Unions
IEE	=	Institution of Electrical Engineers
IGY	=	International Geophysical Year
IISL	=	International Institute of Space Law
ISRO	=	Indian Space Research Organisation
ISS	=	International Space Station
ITU	=	International Telecommunication Union
JAXA	=	Japanese Aerospace Exploration Agency
LACE	=	Lunar Atmospheric Composition Experiment
LDEF	=	Long Duration Exposure Facility
LM	=	Long March
LM	=	Lunar Module
LRL	=	Lunar Receiving Laboratory
LRV	=	Lunar Roving Vehicle
MDA	=	Missile Defense Agency
MEO	=	medium Earth orbits
NAS	=	National Academy of Sciences
NASA	=	National Aeronautics and Space Administration
NERVA	=	Nuclear Engine for Rocket Vehicle Application
NGO	=	nongovernmental organisation
NMD	=	National Missile Defense
NOAA	=	National Oceanic and Atmospheric Administration
NORAD	=	North American Aerospace Defense Command
NPRM	=	notice of proposed rule making
OST	=	Outer Space Treaty
OTV	=	orbital transfer vehicle
PQR	=	planetary quarantine requirements
PSLV	=	Polar Satellite Launch Vehicle

RCC	=	reinforced carbon-carbon
RFI	=	radio-frequency interference
RTG	=	radioisotope thermoelectric generator
SDAG	=	Space Debris Advisory Group
SDI	=	Strategic Defense Initiative
SETI	=	search for extraterrestrial intelligence
SMM	=	Solar Maximum Mission
SOHO	=	Solar and Heliospheric Observatory
SRM	=	solid rocket motor
SSN	=	Space Surveillance Network
STS	=	Space Transportation System
STSC	=	Scientific and Technical Subcommittee
UN	=	United Nations
UNCED	=	United Nations Conference on the Environment and Development
UNCOPUOS	=	United Nations Committee on the Peaceful Uses of Outer Space
UNEP	=	United Nations Environment Program
UNESCO	=	United Nations Educational, Scientific and Cultural Organization
USAF	=	United States Air Force
WHC	=	World Heritage Convention

Chapter 1

SPACE AS A FRONTIER

The exploration of space is often compared with the exploration of the American West in the 19th century, probably because a majority of space writers are American but also because it offers a useful analogy. The thought of the "new frontiersmen" journeying through an unknown and often hostile environment in search of fame, fortune and, ultimately, somewhere to live is eminently transferable to space. So too are the scientific curiosities, unrivalled beauty and undeniable commercial potential of the American West, which drew so many explorers and exploiters alike to the "new frontier."

Another reason that many of those interested in space exploration—at least those exposed to western media—feel comfortable with the thought of space as a frontier is undoubtedly because of the 1960s science fiction series *Star Trek*, which famously labelled space as "The Final Frontier." Whether space, time, or some other concept that we have yet to discover and explore is, in fact, the *final* frontier is open to discussion, but space as a largely unknown and unexplored arena is enough for most of us.

The history of mankind's exploration of this new frontier is a relatively short one in the context of human history—less than half a century—but its influence on mankind has been considerable. For this reason alone, the term describing the period in which space exploration has been possible— the Space Age—deserves its initial capitals, to signify its importance alongside the other "ages of man," such as the Stone Age and the Iron Age. The Space Age began with the launch of Sputnik 1, on 4 October 1957, and will continue, one could argue, for as long as we continue to explore space.

This book is, in part, about our efforts to explore and understand space, but it is also about the space environment itself. The needs of the explorers and the environment have not always been compatible, and the latter has often suffered at the hands of the former. The less positive results of space exploration and development can be seen, for example, in the belts of spacecraft and rocket debris that now linger in orbit around the Earth. The successful growth of space applications, such as tourism, will be dependent on the way we handle this and other problems in the near future.

Space: The Fragile Frontier presents the case for the protection of the space environment in what promises to be an exciting era of space exploration and development. First of all, however, let us consider briefly why we explore space and what space exploration has achieved in its first few decades.

NEED FOR EXPLORATION

Why is space exploration so important? There are many answers to this question, incorporating our continuing quest for land and resources, our unquenchable desire for national and political prestige and our natural affinity towards wealth creation, but the most fundamental answer relates to our needs and desires as human beings.

Exploration is inherent in the human psyche. One has only to observe young children for a short time to confirm this fact. As soon as babies can crawl, they want to know what is in the cupboard, through the door, or around the corner. And once they can walk, they seem to have an in-built desire to investigate the woodshed, the garden, and the world beyond the gate.

Although, for many, the need for exploration and the related power of imagination has been suppressed or subsumed by the time they reach adulthood, others retain the natural curiosity that drove prehistoric man to climb nearby hills and peer into the next valley. It might well be the case that the majority of early men remained content to stay in their own valley, where the food and water supplies were plentiful; indeed, the same philosophy exists in many people's minds today. But there are some who will never be content to remain in the valley, and it is this class of inquisitive human beings that has led the exploration of new continents, climbed high mountains, and braved the extremes of the polar landscapes. The exploration of outer space is simply an extension of that inherent curiosity concerning our environment.

For many of those in the space community—engineers, scientists, and other professionals—space is the ultimate environment for mankind to explore. Although there are still parts of this planet that remain largely unexplored, it is hard to ignore the cliché "been there, done that ..." when one considers exploration on Earth. Much of what we do here now is following in the footsteps of those who have gone before, and it is increasingly difficult to find a new challenge.

Space represents that new challenge, that new frontier.

SHORT HISTORY OF THE SPACE AGE

THE 1950S

The technological and human challenge of space exploration was certainly one of the factors uppermost in people's minds as they prepared to cross the frontier to space in the 1950s. In one sense, the decade was a time of optimism in that science and technology were believed to be the saviors of many of mankind's problems, capable of providing anything from cheap energy and efficient transportation to improved health and welfare. The positive application of technology to space exploration, as opposed to the negative application to warfare, was a part of the optimism and excitement of the times.

Of course, what became known as the "space race" was also politically and militarily driven, principally as a result of the Cold War hostility that existed between the American and Soviet blocs after World War II. This meant that much of postwar rocket research was applied to the development of intercontinental ballistic missiles (ICBMs), rather than rockets for space research (Fig. 1). The goal of the Cold War rocket was to lob the heaviest, most destructive warhead as far as possible into enemy territory, thereby tilting the balance of power in favor of the originator.

As it turned out, it was the Soviets who developed the most powerful rockets. This allowed them to launch much heavier satellites, as shown by the respective weights of the two nations' first satellites. Whereas Sputnik 1 was a 58-cm sphere weighing 83.6 kg, America's first satellite, Explorer 1, was 75 cm long, 15.3 cm in diameter, and weighed about one-tenth that of the Sputnik at just 8.3 kg (Figs. 2 and 3). Moreover, because America took

Fig. 1 Preparing for the Space Age: a Bumper-WAC launch from White Sands Proving Ground, New Mexico, in 1948 (Bumper was a modified version of the German V-2 missile used against the Allies in World War II).

too little notice of announcements by the Soviet Union regarding its intentions to launch a satellite, Explorer 1—launched on 31 January 1958—trailed Sputnik 1 by almost four months.

The launch of that first Sputnik, and the realization that it could as easily have been a warhead, sent shock waves through the American public. They had been led to believe that the Soviets were a relatively backward people, deprived by the communist system of all trappings of high technology. How could they possibly get a satellite into space before America?

Unfortunately for the United States, the respective nations' second satellites did nothing to restore the balance. Sputnik 2, launched on 3 November 1957, weighed over half a tonne (508 kg), whereas Vanguard 1, launched on 17 March 1958, was a tiny 1.5-kg, 16-cm-diam sphere. It was immediately nicknamed the "grapefruit" by Soviet Premier Khrushchev, who could hardly contain his derision. Not surprisingly, this galvanized the competitive urge within America, and the space race was born.

Although it is valid to compare the launch weights of these early space-craft, because satellite mass usually bears a direct relationship to complexity, it is not the whole story. Because American engineers knew their rockets were incapable of launching large payloads, they were forced into a program of miniaturization, which of course continues to this day. Though

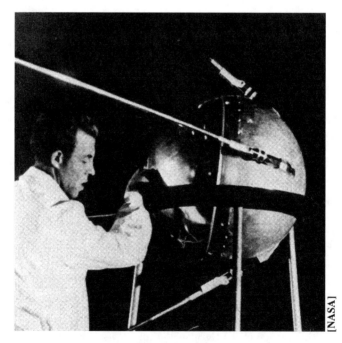

Fig. 2   The first artificial satellite, Sputnik 1.

[NASA]

Fig. 3   A model of America's first satellite, Explorer 1, and its upper stage held aloft by (left to right) William H. Pickering, James A. Van Allen and Wernher von Braun.

small, their early satellites were able to make significant advances in the understanding of the Earth and of the space environment.

The remainder of the 1950s saw an escalation of space launch activity with upgraded versions of the Sputnik, Explorer, and Vanguard satellites either attaining orbit or, equally likely in those early days, crashing back to Earth in a blazing fireball. The most conspicuous successes, when they came, were the Soviet Luna probes aimed at the Moon. Not content with being the first nation to place a satellite in orbit and the first to launch an animal—the dog Laika—as part of the payload of Sputnik 2, the Soviet Union was intent on being the first to the Moon.

Indeed, it is indicative of the desperation of the competing nations to be first in the various heats of the space race that, less than a year after Sputnik 1, both nations were aiming for the Moon. The United States was the first to attempt to launch a spacecraft to the Moon, in August 1958, with a spacecraft in its Pioneer series. Unfortunately, it failed when the Thor–Able launch vehicle's first stage exploded just 77 s into the flight. Close on the heels of its triumphs with the Sputniks, the Soviet Union made its first attempt the following month, but it and two subsequent launches failed even to reach Earth orbit.

Although America's second attempt, in October 1958 with Pioneer 1, failed to reach the Moon, it did set a spaceflight distance record of 113,854 km. The winner of the first heat of the "Moon race," however, was Luna 1 (also called Lunik 1), which was launched on 2 January 1959. It passed within 6000 km of the lunar surface and, in doing so, became the first man-made object to escape from the Earth's gravitational field, a far more important achievement in the annals of spaceflight. In fact, because it continued into an orbit around the Sun, it was sometimes known as Mechta or Planet 10 (there being nine natural, solar planets).

Then, on 13 September 1959, Luna 2 became the first spacecraft to impact the moon. It seems a fairly trivial accomplishment now, but one of the mission's intentions was to deliver a political message in the form of a 26-kg sphere made from tiny medallions decorated with Russia's hammer-and-sickle emblem. Two days later, no doubt to advertise the Soviet Union's triumph, Premier Khrushchev presented President Eisenhower with a replica of one of the pentagonal plates of which the metal sphere was composed.

Although the United States managed a few "firsts" of its own in this period, such as the first photograph of Earth from orbit taken by Explorer 6 in August 1959, space in the 1950s belonged to the Soviet Union (see Table 1). For example, only two months after Explorer 6 reached Earth orbit, Luna 3 returned the first image of the Moon's far side. This was a technopolitical coup akin to discovering the lost city of Atlantis, because, until then, one could only imagine how the lunar far side would appear. Because the Moon is in captured rotation about the Earth, it presents only the familiar face to the Earth at all times. The pictures showed that there were many more craters and fewer lunar seas, or maria, than on the near side. Perhaps more importantly, they gave an inkling of the advantage to science of sending spacecraft out into the solar system.

THE 1960s

Despite the technical and political triumphs in space recorded in the late 1950s, the space race had only just begun. The recognized goal of the 1960s was to launch the first man into space. It was one thing to send dogs, mice, and monkeys to experience space on our behalf, but the final frontier would only truly be open for exploration if man were personally involved.

This was the goal of the Soviet Vostok and American Mercury programs, the respective spacecraft each being designed to carry a single cosmonaut or astronaut. As history has recorded, Yuri Gagarin (Fig. 4) became the first man in space on 12 April 1961, followed closely, on 5 May, by Alan Shepard. However, although America appeared to be gaining on the Soviet Union, while Shepard's Mercury capsule executed a suborbital hop on a mission lasting 15 minutes and 22 seconds, Gagarin's Vostok actually orbited the Earth for 1 hour and 48 minutes at a greater overall altitude. (John Glenn became the first American in orbit in February 1962.)

Beaten, yet again, by the Soviet Union and suffering a politically disastrous defeat in the Cuban Bay of Pigs confrontation, the United States needed a boost to its collective morale. Thus it was a combination of events that led President John F. Kennedy to make one of the most famous political directives of all time. On 25 May 1961, to a packed U.S. Congress, he declared: "I believe that this Nation should commit itself to achieving the goal, before this decade is out, of landing a man on the Moon and returning him safely to the Earth."

It still seems incredible today that, when Kennedy made his commitment, the United States had little more than 15 minutes of manned spaceflight experience, accrued on a flight that reached an altitude of 187 km (about the length of Long Island, New York). The Moon, by contrast, was about 375,000 km away, and a successful lunar mission would last at least eight days. The enormity of the engineering challenge was unprecedented.

However, the political decision, and the funding allocations to support what became the Apollo program (Fig. 5), provided a spur to the development of space technology that led to the realization of the goal in July 1969. It is significant not only in the scope of the history of technology, but of history itself, that only eight years after Yuri Gagarin became the first human being to orbit the Earth, and less than 12 years after Sputnik 1, two men were standing on the surface of the Moon.

Although the Soviet Union had its own program to land men on the moon, all four launch attempts of its N-1 Moon rocket, made between February 1969 and November 1972, were failures, and the program was cancelled. The space race, and the Moon race within it, was finally over.

At the time, however, with the Soviet Union adding the first woman in space, first multiperson crew, and first spacewalk to its score card, it was easy to forget that there was more to space exploration than the high-profile manned missions. The 1960s also saw the first astronomical observatory in space (NASA's Orbiting Solar Observatory, OSO-1, launched in March 1962); the first spacecraft to land, as opposed to crash, on the Moon (Luna 9 in January 1966); and the first spacecraft to reach the vicinity of Venus and Mars (Venera 1 in February 1961 and Mars 1 in November 1962). The first successful flybys of Venus and Mars were made,

Table 1   Summary of selected space firsts

Launch date	Spacecraft	Nation	Event
4 October 1957	Sputnik 1	USSR	First satellite in space (Earth orbit)
3 November 1957	Sputnik 2	USSR	First animal in space (dog "Laika," Earth orbit)
2 January 1959	Luna 1	USSR	First spacecraft to escape Earth's gravitational field
7 August 1959	Explorer 6	USA	First photograph of Earth from orbit
12 September 1959	Luna 2	USSR	First spacecraft to impact Moon (13 September 1959)
4 October 1959	Luna 3	USSR	First image of lunar far side
1 April 1960	Tiros 1	USA	First weather satellite
12 August 1960	Echo 1	USA	First passive communications satellite (reflective balloon)
4 October 1960	Courier 1B	USA	First active communications satellite (radio repeater)
12 February 1961	Venera 1	USSR	First spacecraft to reach Venus (inactive)
12 April 1961	Vostok 1	USSR	First man in space (Yuri Gagarin, Earth orbit)
7 March 1962	OSO-1	USA	First astronomical observatory in space (Orbiting Solar Observatory)
10 July 1962	Telstar 1	USA	First privately owned communications satellite (low Earth orbit)
27 August 1962	Mariner 2	USA	First successful flyby of Venus (14 December 1962)
1 November 1962	Mars 1	USSR	First spacecraft to reach Mars (inactive)
16 June 1963	Vostok 6	USSR	First woman in space (Valentina Tereshkova, Earth orbit)
19 August 1964	Syncom 3	USA	First spacecraft to attain geostationary orbit (communications satellite)
12 October 1964	Voskhod 1	USSR	First multiperson crew (three cosmonauts on 24-h mission)
28 November 1964	Mariner 4	USA	First successful flyby of Mars (14 July 1965)
18 March 1965	Voskhod 2	USSR	First EVA in Earth orbit (Alexei Leonov)[a]
31 January 1966	Luna 9	USSR	First spacecraft to land on Moon (hard landing on 3 February 1966)

(Continued)

Table 1 Summary of selected space firsts (continued)

Launch date	Spacecraft	Nation	Event
31 March 1966	Luna 10	USSR	First spacecraft in lunar orbit (3 April 1966)
30 May 1966	Surveyor 1	USA	First spacecraft to soft land on Moon (2 June 1966)
21 December 1968	Apollo 8	USA	First manned spacecraft in lunar orbit
16 July 1969	Apollo 11	USA	First manned spacecraft on the Moon (Eagle lunar module, touchdown 20:17 hrs GMT 20 July 1969)
16 July 1969	Apollo 11	USA	First EVA on the Moon (Neil Armstrong, boot down 02:56 h GMT 21 July)[a]
17 August 1970	Venera 7	USSR	First radio transmission from surface of another planet, Venus (landed 15 December 1970)
12 September 1970	Luna 16	USSR	First return of lunar samples to Earth (24 September 1970)
10 November 1970	Luna 17	USSR	First teleoperated rover on Moon (Lunokhod 1, landed 17 November 1970)
19 April 1971	Salyut 1	USSR	First space station (Earth orbit)
2 March 1972	Pioneer 10	USA	First spacecraft to escape solar system (passed orbit of Pluto 13 June 1983)
20 August 1975	Viking 1	USA	First spacecraft to soft land on Mars (20 July 1976)
12 April 1981	Columbia (mission STS-1)	USA	First space shuttle/partially reusable launch vehicle
2 July 1985	Giotto	ESA[b]	First close encounter of a comet (Comet Halley, 13 March 1986)
4 December 1996	Mars Pathfinder	USA	First teleoperated rover on Mars (Sojourner, touchdown 4 July 1997)
28 April 2001	Soyuz to ISS	USSR	First space tourist (Dennis Tito to International Space Station)
4 October 2004	SpaceShipOne	USA	First privately built spacecraft to exceed an altitude of 100 km twice in succession

[a]EVA = extravehicular activity. [b]ESA = European Space Agency.

[NASA/Asif Siddiqi]

Fig. 4   Yuri Gagarin in the bus to the launch pad on 12 April 1961. Seated behind him is his back-up, Gherman Titov, while cosmonauts Grigoriy Nelyubov and Andrian Nikolayev stand.

respectively, by NASA's Mariner 2 on 14 December 1962 and Mariner 4 on 14 July 1965.

Moreover, space applications were expanding beyond the purely scientific realm, and the 1960s also saw the launch of the first weather satellite, Tiros 1, in April 1960 and that of the first privately owned (non-government) communications satellite, Telstar 1, in July 1962. This was the beginning of a new era in the Space Age, the era of commercial development. In the same way that mankind has always found ways to exploit resources, ostensibly for the betterment of all involved, the newly discovered resources of Earth orbit were now being developed.

The advantages were evident to the pioneers. Meteorological satellites would enable better weather forecasts and early warning of severe conditions that would allow lives to be saved. In fact, since 1966, the entire Earth has been photographed at least once a day, and no tropical storm has escaped detection and daily tracking. Meanwhile, communications satellites would enable isolated settlements to join what became known as the global village, without the need to string cables from poles across hundreds of kilometers of desert or, in the case of the Philippines, link some 13,000 islands by undersea cable!

[NASA]

**Fig. 5   Apollo 12 astronaut Al Bean descends towards the Moon's
surface on the second manned lunar mission.**

Despite the technological advances made through space exploration in the
1960s, especially those which allowed men from Earth to stand on the
surface of another planetary body, perhaps the most significant legacy of
the Space Age is the image of the Earth rising above the surface of the
Moon (see Fig. 6). For the first time, it placed the Earth in perspective
with the rest of the universe, showing it as insignificant and fragile, sur-
rounded by a sea of space. As such, it served to highlight the precarious
position of its inhabitants and their environment and is often cited as an inspi-
ration for the Green movement.

THE 1970s

In a sense, the 1970s was as much a part of the Apollo era as the 1960s:
following the Apollo 11 and 12 missions in 1969, there was one in 1970
(the infamous Apollo 13) and two in each of the subsequent years. Then,
after just six manned landing missions, it was all over. Public apathy
born of repeat showings on television and political resistance to NASA's
ever-increasing funding requests forced the cancellation of the final three

Fig. 6 Earthrise from Apollo 8 in lunar orbit (the horizon is about 780 km from the spacecraft and the width of the image at the horizon is about 175 km).

missions (Apollo 18, 19, and 20) and the termination of manned lunar exploration.

However, good use was put to some of the excess hardware in the guise of the Skylab space station, manufactured from one of the surplus third stages of the Saturn V moon rocket (see Fig. 7). The station was launched by a Saturn V on 14 May 1973 and played host to three crews of three astronauts, who visited for periods of 28, 59, and 84 days, in the months that followed. The program was a great success, proving first of all that astronauts could repair the station, which had been damaged during launch, and survive in space for much longer periods than had previously been possible. In addition to conducting research into how human beings coped in the space environment, they helped to make significant progress in Earth observation and solar astrophysics. In many ways, Skylab was a precursor to the Shuttle-borne Spacelab module and the International Space Station of later decades, but, like the Apollo program that sired it, it was curtailed at its height, and the station was vacated in February 1974.

The Soviet Union, by contrast, was more committed to the concept of the manned space station, to which it had applied itself following its abortive attempts at lunar exploration. Its first space station, Salyut 1, was launched in April 1971, and was followed throughout the 1970s and 1980s by a series of similar stations dedicated to either military or civil applications.

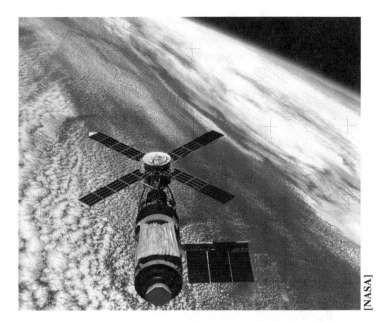

[NASA]

Fig. 7   The Skylab space station in low Earth orbit.

For the majority of the 1970s, therefore, the Soviet Union ruled low Earth orbit as far as manned flights were concerned.

The United States conducted only one manned flight between the end of the Skylab program in 1974 and the first Space Shuttle mission in 1981: the Apollo–Soyuz Test Project (ASTP). It was another largely politically driven mission, which developed from an agreement between President Nixon and Premier Kosygin, an attempt at détente in the midst of the continuing Cold War. The mission featured a docking between an American Apollo and a Soviet Soyuz spacecraft on 17 July 1975 and numerous handshakes among the three astronauts and two cosmonauts. Although it was a brave attempt to get the two superpowers talking, its only lasting impact was felt by those personally involved; the mission, and the feelings of détente it engendered, were soon forgotten.

Beyond the fickle eye of the media, however, the expansion of the new frontier was progressing nicely. A space industry that had developed from the existing aerospace and electronics sectors in the 1960s was reaching a sort of adolescent stage, in which it was able to offer its customers a wide range of spacecraft and supporting equipment. Satellites for communications, meteorology, remote sensing—and of course their equivalents in the secret world of government and military applications—were being designed, built, and launched in increasing numbers.

International cooperatives for satellite communications and meteorology were reaching maturity, and the use of their satellites and the data they provided was becoming integrated into society. TV viewers around the world were becoming used to seeing "live via satellite" on their screens, and meteorologists were beginning to wonder how their predecessors had managed without satellite images.

Meanwhile, an American remote-sensing satellite, known as Landsat, was returning images of the Earth of a type never seen before. Landsat 1, launched in July 1972 and formerly known as the Earth Resources Technology Satellite (ERTS), was designed primarily for land-use studies, but also produced images of coastal areas, rivers, towns, and cities. Although the resolution of its sensors was insufficient to make out anything but the most significant man-made structures, the largest buildings, airports, and major highways were visible. More importantly, the satellite's infrared sensors could distinguish healthy crops from unhealthy ones and detect the boundaries between different types of soil, rock, and other materials. It was the beginning of a scientific, and later commercial, revolution in space imaging.

There was also a revolution going on in planetary exploration. Although NASA had been forced to take a backseat in manned spaceflight, it was doing no such thing in unmanned, robotic, exploration. For example, in March 1972 it launched Pioneer 10 to Jupiter, the largest planet of the solar system. The spacecraft made a successful close approach of 131,400 km in December 1973, returning fascinating close-up images of the planet's turbulent atmosphere, including the famous Great Red Spot. Pioneer 10 was followed by Pioneer 11, which made a similar flyby of Saturn, and later in the decade by a series of two Voyager spacecraft, which eventually visited Jupiter, Saturn, Uranus, and Neptune, further expanding our knowledge of these distant planets (Fig. 8).

Nor were the closer planets ignored in this decade of planetary exploration. In 1975, for example, NASA launched two Viking spacecraft, each comprising an orbiter and a lander, to Mars. The following year, the landers became the first two spacecraft to execute soft landings on the red planet, returning stunning images of a rock-strewn world (Fig. 9). Although their onboard laboratories failed to find conclusive evidence for life on Mars, their data continued to be the subject of intense scientific discussion for the following two decades.

Although the space race was less intense than it had been in the 1960s, it had simply shifted down a gear, and there was still a culture of competition between the United States and the Soviet Union. Thus, while the United States had found success with Mars, the Soviet Union had made its mark on the inhospitable surface of Venus. Indeed, its Venera 7 probe had made the first radio transmissions from the surface of another planet back in December 1970. And closer to home, the Soviets had also succeeded in

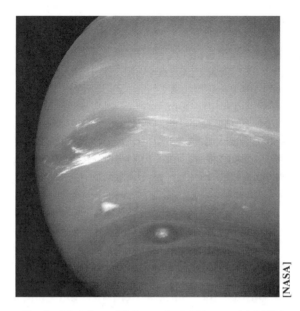

Fig. 8   The planet Neptune from Voyager 2 in 1986.

landing two Lunokhod teleoperated rovers on the Moon—the first to operate on a planetary body other than the Earth—and in returning lunar samples to the Earth using an automated spacecraft.

By the end of the 1970s, the unmanned exploration of the solar system was well underway, and, back on Earth, the development of a commercial space industry had reached the point where it was no longer simply an adjunct to

Fig. 9   The Viking 2 lander on Mars in 1976.

the respective companies' aircraft or weapons divisions. Space systems were products in their own right.

Nor was space exploration and development confined to the superpowers. Through the European Space Agency (ESA), Europe had developed the skills to build satellites and launch vehicles, which, among other things, had produced the world's first commercial satellite launcher, Ariane. Prior to Ariane, the first flight of which occurred in December 1979, space launch vehicles had been developed from existing ICBMs; Ariane was designed from the outset as a commercial vehicle and rose to become the field's market leader in the 1990s.

THE 1980s

Whatever else happened in space in the 1980s—and there was plenty going on—it was the decade of the Space Shuttle, a launch system designed to reduce the cost of access to space. As it turned out, the process of launching and refurbishing the Shuttle system was more difficult and expensive than first imagined, but it remains the nearest thing we have to what science-fiction aficionados might consider a real spaceship (Fig. 10).

The Shuttle orbiters themselves, as opposed to the solid rocket boosters and external propellant tank which completed the Shuttle launch system, were the world's first reusable spacecraft, each designed to fly at least 100 missions. Following a long period of development in the 1970s, the first mission was flown on 12 April 1981 by Columbia, one of a fleet of four spacecraft—it was 20 years to the day since Yuri Gagarin had orbited the Earth. An indication that the reusable spacecraft concept had merit was provided by the fact that the Soviet Union developed its own space shuttle, named Buran, which bore a striking resemblance to the U.S. Shuttle. It was launched on an unmanned test flight on 15 November 1988, but, as a result of political uncertainty surrounding the demise of the Soviet Union, it was its first and only flight.

One of the worst accidents in the history of manned spaceflight occurred just five years into the Shuttle program when, on 28 January 1986, the Challenger exploded, killing its crew of seven. The cause was traced to the failure of a pressure seal on one of the solid rocket boosters, which had failed because the Shuttle had been launched in excessively cold weather. This placed a hold on Shuttle launches of more than two and a half years while the technical and safety problems were addressed and a replacement orbiter was built. Following the accident, which occurred on the 25th mission, the Shuttle system flew more than three times that number of successful missions before its reputation was tarnished once again by the destruction of Columbia, on the fleet's 113th flight on 1 February 2003.

[NASA]

Fig. 10   A launch of NASA's Space Shuttle.

Coincidentally, the month after the Challenger accident, the Soviet Union launched the core module of its Mir space station. It was similar to the Salyut 7 space station, which had been launched in April 1982, and had a number of docking ports to accept visiting crews and resupply spacecraft. An astrophysics module called Kvant was added in April 1987, followed by Kvant 2 in 1989 and a further three modules in the 1990s (Fig. 11). During its lifetime, Mir provided invaluable, long-term experience of the space environment and received many international visitors including, later, U.S. Space Shuttle crews. All in all, it was one of the most successful space programs in the short history of space exploration.

Beyond the field of manned spaceflight, the satellite industry also reached a degree of maturity. Communications satellites had become a recognized part of the global telecommunications network, and many nations had procured their own fleets of satellites for domestic broadcasting and telecom services. Some leading nations also owned remote-sensing and weather satellites and had even developed domestic military communications systems. All in all, the space industry was optimistic about the

[NASA]

Fig. 11   Russia's Mir space station.

future, predicting the need for larger and larger satellites and more ambitious missions and applications, not the least those that would monitor the oceans, the ozone hole, and the Earth's climate in general.

Although planetary exploration took something of a backseat in the 1980s, by the end of the decade mankind had succeeded in visiting (albeit remotely) all the planets except Pluto. Significantly, in terms of pushing the boundaries of the space frontier, Pioneer 10 became the first spacecraft to escape the solar system by passing the orbit of Pluto in June 1983.

There were also a number of unusual missions, including the first close encounter of a comet made by ESA's Giotto spacecraft (Fig. 12) on 13 March 1986: traveling at a relative velocity of 68 km/s, it passed within 600 km of Halley's Comet, taking the first close-up images of a cometary nucleus. It was later retargeted towards Comet Grigg-Skjellerup and made a close approach of 200 km on 10 July 1992.

THE 1990s

The 1990s was a period of some turmoil and great change within the space sector, in part because the self-confidence of the previous decade had encouraged prospective satellite operators to believe their own hype. One of the prime examples was the Iridium system for satellite-based mobile telephone services: having spent over four billion dollars establishing a constellation of 66 satellites in low Earth orbit (LEO), Iridium's backers found that too few customers wanted to subscribe to the service, and the company was eventually declared bankrupt. The reasons behind Iridium's demise were many and varied, but included failing to make the service competitive—in terms of both cost and capability—with fast-developing terrestrial mobile systems.

So important was Iridium's business failure that the subsequent slump in the market for constellation systems and satellite services in general was termed "the Iridium effect." Coupled with a number of regional recessions, a global telecommunications slump and the failure of a number of so-called dot-com enterprises, the satellite industry's own deficiencies made the end of the 1990s, particularly, a very difficult time for space commerce.

The decade also started badly for space science, when it was discovered, following the launch of the Hubble Space Telescope (HST) in April 1990, that its main mirror suffered a spherical aberration causing its images to be

[EADS/British Aerospace]

Fig. 12 ESA's Giotto Comet Halley interceptor (note sacrificial shield at bottom of spacecraft).

blurred. An investigation showed that the mirror had been ground incorrectly as a result of a measurement error. It is a testament to the ingenuity of space engineers, and to the capabilities of the Space Shuttle and its astronauts, that the HST, shown in Fig. 13, is no longer remembered mainly for its blurred images. A servicing mission in 1993 and subsequent upgrade missions have allowed the telescope to make many successful observations, including the derivation of a new value for the Hubble constant (the constant of proportionality between the recession speeds of galaxies and their distances from each other) and thus an improved estimate of the age of the universe. Apart from that, some of the images returned have been nothing short of stunning.

Another mission that captured the public imagination was Mars Pathfinder, the spacecraft that broke the 20-year, post-Viking hiatus by conducting an unusual balloon-cushioned landing on Mars on 4 July 1997. It carried a teleoperated rover, called Sojourner, which became one of the best known spacecraft ever, because of the rising popularity of the Internet. As Sojourner trundled about the surface, NASA's website received an unprecedented number of hits as people around the world clamored to follow the little rover's meanderings.

Meanwhile, on 17 February 1998, Voyager 1 overtook Pioneer 10 to become the most distant of mankind's creations: it was 70 astronomical units (AU)—some 10.4 billion km—from Earth. (1 AU is equivalent to the distance between the Earth and Sun: 93 million miles, or 148 million km.)

Fig. 13   The deployment of NASA's Hubble Space Telescope in low Earth orbit.

Another major space success of the 1990s was the Global Positioning System (GPS), as indicated by the fact that the acronym has entered the language and that GPS receivers are used by yachtsmen, motorists, and walkers on a daily basis. In fact, its use by civilians around the world is a bonus because the system was designed primarily for the U.S. military. Its capabilities were proven to great effect during the 1990 Gulf War, which, as a result of its use of GPS and communications satellites, was termed the first space war. Perhaps the most telling illustration of the satellite's capabilities was provided by television images of GPS-guided cruise missiles flying down streets in Baghdad before hitting individual buildings "bang-on-target."

In retrospect, however, the decade should be remembered for the world's largest ever program of international space cooperation, the International Space Station (ISS). Its origins date back to 1984, when U.S. President Ronald Reagan directed NASA to build a space station. However, redesigns and cost overruns delayed the beginning of orbital assembly until November 1998, by which time (as a result of the dissolution of the Soviet Union) Russia had joined Europe, Canada, and Japan as a partner.

Surprisingly, considering the ISS had been born in the United States, the first module to reach orbit was the Russian-built Zarya, formally known as the functional cargo block (or, from its Russian name, FGB). The space race had effectively gone full circle: Russia had won the race to launch the first satellite and first man into space, lost the race to the Moon, and was now providing, under contract to NASA, the first module for an American-led space station.

Unfortunately, although the Mir space station secured Russia's domination of LEO well into the 1990s, the nation could no longer afford to operate its own station in parallel with the ISS, and Mir was deorbited in March 2001.

SPACE IN THE 21ST CENTURY

The space industry made the transition to the new millennium with a degree of uncertainty, brought about by global investment problems and the knowledge that there were too many space-related companies serving too few customers.

However, the continuing integration of satellite services into everyday life provided a degree of confidence in the long-term development of space commerce. Communications satellites continued to provide Internet-related services, broadcasting satellites were expanding their remit from television to radio, GPS receivers continued to infiltrate themselves into society, and high-resolution commercial imaging satellites began to deliver image resolutions better than 1 m.

[NASA]

Fig. 14   The International Space Station (ISS).

Meanwhile, plans were being made to send an unprecedented number of unmanned spacecraft to Mars, and the ISS continued to grow in low Earth orbit (Fig. 14). A highlight of its early life for space explorers, vicarious and actual, was a visit by the first space tourist, Dennis Tito, in April 2001. He had originally signed up for a trip to Mir, but because it was no longer in orbit his reported $20 million ticket had been transferred to the ISS. Although the price will have to drop substantially before space tourism becomes as affordable as adventure tourism here on Earth, Tito's visit set a precedent, and potential tourists began queuing up to visit the ISS.

Another right of passage in the field of space exploration was marked in October 2003, when the People's Republic of China became the third nation to place one of its citizens in orbit using its own technology. The launch of China's first astronaut, Yang Liwei, in the Shenzhou 5 spacecraft, provided another perspective on the future development of the final frontier, with some pundits even predicting that the first man on Mars could be Chinese.

[NASA]

Fig. 15   Mars—the new frontier? Note the Valles Marineris canyon system across the middle of the image and the massive volcanoes of the Tharsis range at far left.

It would not be unreasonable to suggest that U.S. President George Bush had this possibility in mind when he announced, in January 2004, that Americans would return to the Moon and then explore Mars (Fig. 15). Whatever the impetus behind this so-called "vision," the design of the first flag to be planted on the red planet remains open for speculation.

EXPLORE ... DEVELOP

Predicting the future is a difficult business at the best of times, and predicting developments in space technology is no easier than in any other field of human endeavor. As recently as 1990, few could have predicted, for example, that Russia would be a leading partner in an international space station, or that commercial GPS receivers and handheld satellite telephones would be available by the end of the decade.

When the original *Star Trek* television series was made in the mid-1960s, the majority of satellite earth stations used parabolic antennas measuring some 30 m in diameter, largely because the signals transmitted by satellites were so weak. The concept of communicating with a spacecraft in a stationary orbit, such as the starship *Enterprise*, using a device the size and weight of an electric shaver was, without doubt, pure fiction. Who could have predicted that this 23rd-century device would become reality before the end of the 20th century?

Of course, not all developments in space technology occur sooner than anticipated by science fiction or even by science fact: the Mars missions, Moon bases, and solar-power satellites predicted in the 1960s have not materialized, mainly because of the cost. As developments have shown, in the real-world space activities are inseparable from other human activities, such as financial speculation, politics, and war. In the long term, this is good for space exploration and development because space-based assets are useful to mankind. If space were simply an expensive hobby, cultural fad, or political whim, things would be very different.

Luckily for space advocates, there are practical reasons for space exploration and development, which have served to keep the frontier open. The exploration of space has already led to its development as a human resource. Although most people take communications and weather satellites entirely for granted, the world would be a very different place without them: no satellite television, no satellite weather maps, no live news reports from faraway places. Of course, lifelong valley dwellers would argue that they managed just fine before all this newfangled technology came along. And they would be right, but then their forebears managed before the invention of the telephone, the electric water heater, and the flush toilet.

Today, there are thousands of people around the world who owe their lives to satellites because they warned them to shelter from a hurricane, allowed them to call for rescue services, or simply told them—in the case of GPS—which way to go home. Had they stayed at home, they might not have needed the satellites, but many members of the human race are not content to stay at home; they have an inherent need to explore.

The alternative—remaining on Earth or in near-Earth orbit—is not a viable option for the human race because it denies this need. Moreover, any global decision to call a halt to space exploration when we have come so far would be tantamount to technological and psychological stagnation: the equivalent of pulling the covers over our heads and telling the universe to go away. Having opened the door to space exploration, mankind as a race will find it very difficult to close it again. To do so would imply failure, contentment with the status quo, and a diminution of the spirit that makes the human race what it is. It could even mark the beginning of a slow descent into a technological dark age where introspection and conservatism become the norm.

Many would say that this places space exploration on too high a pedestal in the context of mankind's activities, but let us return to the analogy of the American West. If the early explorers, and subsequently settlers, had remained in the East, content with their tranquil valley, would California be the rich and populous state it is today? Possibly, but it would not be a major contributor to the U.S. Internal Revenue Service. Those early attempts at exploration, settlement, and development certainly paid off.

Completing the analogy means that, sooner or later, space exploration will evolve into true space development, in effect an industrialization of space. This could have many manifestations, from the in-orbit manufacturing of spacecraft and space habitats to the production of high-value pharmaceuticals and semiconductor crystals. It could include a tourism industry, a lunar mining industry, and a supporting space transportation industry.

Among other developments, the beginnings of space tourism on the ISS and the rocket delivery of cremated human remains to orbit suggest that, some time in the 21st century, the space environment will become an extension of our terrestrial business and domestic environment. The completion of the Ansari X-Prize competition in late 2004, which saw the launch of a privately built spacecraft on a suborbital trajectory, was a further sign that the floodgates of space tourism might be about to open. This otherwise welcome development will bring its own problems... to which solutions will have to be found.

One of the prerequisites for the development of an environment and the use of all it has to offer is surely an understanding or appreciation of that environment. Unfortunately, at present, the human race has an inadequate understanding and appreciation of the space environment, and this is a problem that needs to be solved before explorers become settlers of the final frontier.

Chapter 2

SPACE AS AN ENVIRONMENT

To say that the space environment is not well understood is an understatement. Serious scientific investigation of space has been underway for less than 50 years, and only a few hundred people have visited the shallow sphere of space we call low Earth orbit. Just 27 men have ventured beyond Earth orbit to the Moon, a mere 300,000 km distant, and only four unmanned spacecraft have left the confines of the solar system.

So, how should we define the space environment? As with all definitions—and certainly with definitions of the Earth's environment—it depends on one's viewpoint. For example, from a geomorphological perspective the Earth can be said to have many different environments: polar, tropical, and temperate; highland, lowland, and subterranean; land, sea, and air. Likewise, from a social perspective, different areas of the Earth might be described as rural or urban; natural or manmade; uninhabited or overpopulated.

However, if we confine ourselves to simple definitions of the word environment, which include "surroundings" and "external conditions," we find that these are readily transferable to space.

SURROUNDINGS

In simple, admittedly geocentric terms, the space environment can be defined as that part of the universe beyond the Earth's atmosphere (because the atmosphere is considered part of the Earth's environment). Of course, in universal terms, the Earth is just another planetary body and is therefore part of the space environment. That aside, the geocentric definition of the space environment depends on where the atmosphere ends and space begins.

To assist that decision, it might be helpful to remind ourselves of the structure of the atmosphere, which is divided into several layers or regions called spheres, the boundaries between which are known as pauses (a nomenclature introduced by British mathematician and geophysicist Sydney Chapman). They are as follows:[1]

    1) Troposphere: from the surface to the tropopause (at an altitude of 10–12 km, dependent on latitude).

2) Stratosphere: from the tropopause to the stratopause (at about 50 km).
3) Mesosphere: from the stratopause to the mesopause (at about 85 km).
4) Thermosphere: from the mesopause to about 600 km.
5) Exosphere: above about 600 km, where individual atoms can reach escape velocity without collisions.

The stratosphere and mesosphere are, together, often termed the middle atmosphere; the thermosphere and exosphere are generally known as the upper atmosphere, but also include the lower reaches of the space environment (Fig. 16).

Of course, most of our day-to-day lives are conducted within the troposphere, and even commercial, subsonic aircraft fly below the tropopause. Only supersonic aircraft and high-altitude balloons can fly within the stratosphere, while rockets punch straight through it on their way to space. Although the X-15 rocket-propelled aircraft of the 1960s reached record

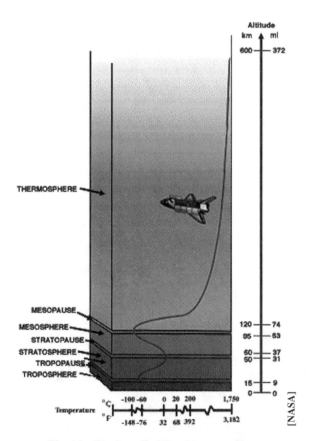

Fig. 16   The layers of Earth's atmosphere.

altitudes of about 108 km, well above the mesopause, the thermosphere tends to be reserved for relatively short-lived Earth observation and reconnaissance satellites and manned space stations. Most satellites are stationed in the exosphere, which, for all extents and purposes, can be considered as space.

Perhaps surprisingly, there is no categorical agreement on where space begins, although it is generally understood to start at about 100 km above sea level (somewhere in the lower thermosphere). In the early years of the Space Age, 50 miles (80.5 km) was a convenient figure to use, and it was this definition that allowed many of the X-15 pilots to earn their astronaut wings (at least in the eyes of the U.S. Air Force). In 2004, however, the legitimacy of the 100-km limit was supported by the completion of the Ansari X-Prize competition to launch a privately built spacecraft above this altitude twice in succession. As a result, it seems likely that this figure, which is recognized by the Fédération Aéronautique Internationale (FAI) for record-certification purposes, will become the *de facto* boundary to the space environment.

A more scientific reason for choosing 100 km is that spacecraft cannot complete an orbit below this altitude because friction with the atmosphere causes them to reenter. (For comparison, satellites in the upper thermosphere will remain in orbit for at least a decade.) One could therefore attempt to define the boundary of space by reference to atmospheric density, but the altitude of this boundary would vary with time because the atmosphere expands when solar activity is high.

This problem of boundary definition is not new, of course. A similar difficulty was faced when attempting to define the boundaries of nations bordering an ocean. In this case, it was solved by defining a standard, measured limit beyond which the ocean was defined as international waters. By contrast, despite the growing acceptance of the 100-km limit, there is still no internationally agreed, or legally binding, definition regarding the boundary between Earth and space.

EXTERNAL CONDITIONS

There is, however, far more agreement on the environmental conditions found beyond the Earth, because they have been under continual study since the beginning of the Space Age. In fact, the launches of the first satellites occurred within an important period of scientific investigation known as the International Geophysical Year (IGY).

In the late 1950s, when plans for the IGY were being developed, outer space was seen as an exciting extension to the geophysical environment, and, because the technology to launch a satellite was becoming available, it seemed logical to include this aspect of investigation. Indeed, the IGY organizing committee, which met in Rome in 1954, made the following recommendation:

> In view of the great importance of observations during extended periods of time of extraterrestrial radiations and geophysical phenomena in the upper atmosphere, and in view of the advanced state of present rocket techniques, the Special Committee for the International Geophysical Year recommends that thought be given to the launching of small satellite vehicles.[2]

The phrase "extended periods of time" is significant in that, before satellites were available, observation times were limited to the few minutes available from sounding rockets. As it turned out, the satellite program chosen by U.S. President Eisenhower to contribute to the IGY was Vanguard, an essentially civilian program organized under the auspices of the U.S. Navy. As mentioned in the preceding chapter, Vanguard was America's second satellite and the world's fourth to reach orbit, following Sputnik 1, Sputnik 2, and Explorer 1.

Because Sputnik 1's main technical function was to transmit radio signals, chosen primarily to get the satellite noticed by radio amateurs as well as professionals, it was not designed with science in mind. However, it was possible to convey data on the internal temperature of the spacecraft by varying the duration of the signal pulses and the pauses between them. This was, in effect, the first attempt at spacecraft telemetry. Moreover, it also allowed a degree of serious research into the propagation of radio signals through the atmosphere, which would earn it the honor of being the first orbiting space science experiment.

As an indication of how much information could be gained from so little hardware—at least in the days of relative ignorance of space and atmospheric science—Sputnik 1's carefully machined and highly polished spherical surface allowed an additional line of research. Measurements of the differences between the optical and radio "rising and setting" of the satellite made it possible to determine the distortion of the radio beam by the ionosphere and to deduce its electron content.[3]

Sputnik 2 carried a package of solar ultraviolet and X-ray detectors, which made it, in a sense, the first space astronomy satellite. This was the first time it had been possible to conduct relatively long-term astronomical observations from orbit. In fact, to complement the suite of onboard instruments, a cosmic ray experiment was mounted on the body of the launching rocket itself. Of course, it was also the first space biology/life-support experiment, by virtue of its canine passenger, Laika.

Although the United States' first satellite, Explorer 1, was one-tenth the mass of Sputnik 1, it carried a far more advanced payload. For example, it incorporated thermometers to record internal temperatures, which varied between 10 and 40°C, and others to measure external temperatures, which varied between +60 and −30°C depending on the Sun's illumination of the satellite. The payload also included a microphone to detect meteor impacts (about one every 15 days), a cosmic ray detector, and a Geiger counter—the most significant part of the payload because it was instrumental in discovering what became known as the Van Allen radiation belts.

Likewise, Vanguard 1 was significant, despite its small size, in being the first satellite to derive its power from solar cells (Fig. 17), the descendants of which now power the majority of spacecraft, including the International Space Station. Although its payload was severely limited in scope, tracking the changes and perturbations of its orbit led to important conclusions on the shape of the Earth and produced an accurate figure for the flattening of the poles. The satellite Khrushchev had called "grapefruit" also showed that the southern hemisphere was larger than the north, producing the journalistic witticism "Grapefruit says Earth is Pear-shaped."[4]

Since then, of course, a great deal more has been discovered about conditions in the space environment, and much has been written on the subject. For our purposes, suffice it to say, in summary, that the space environment constitutes the physical environment of space, which includes its temperature, radiation, and gravitational properties. By extension,

[NASA]

Fig. 17 America's second satellite, Vanguard 1, was just 16 cm in diameter and weighed 1.5 kg (about the size of a grapefruit).

because the planetary bodies and other constituted matter, such as comets and meteors, exist within this space, they too are part of the space environment.

To be specific, and to confine our interest to the solar system, the space environment includes the Sun; the nine recognized planets of the solar system (Mercury, Venus, Earth, Mars, Jupiter, Saturn, Uranus, Neptune, and Pluto); their natural satellites or moons; and everything else in between (Fig. 18). This, however, is just the "boy's own" definition of the solar system. Scientifically speaking, the space environment also includes the products and influences of these bodies.

The Sun's influence, for example, extends far outside the orbit of Pluto to the heliopause, the boundary between the heliosphere (the region of space influenced by the charged particle flux from the Sun), and interstellar space. In mid-2005, it was announced that Voyager 1 was thought to be close to this notional outer boundary of the solar system, and more definite results were expected by the end of the decade.

The Earth's influence, to take another example, is manifested in its gravitational attraction, which holds the Moon and other satellites in their orbits, and in its reflective properties, which produce earthlight in the lunar night. And the influence of Jupiter on *its* moon, Io, is significant because its gravity provides the engine for Io's volcanic activity, making it one of the most active bodies of the solar system.

From a scientific point of view, the tiny part of the space environment that lies within the solar system is a fascinating thing to study. At the same time, from the viewpoint of those with interests in developing space, it is also resplendent with resources.

ORBITAL RESOURCE

Currently, the most important assets or resources of space are the orbits around the Earth, because they are home to most of our satellites (for civil communications, meteorology, Earth observation, astronomy, and military applications) and to all of our manned missions. Theoretically, there are any number of different types of orbit, with different altitudes, inclinations and eccentricities, but in practice we use only a limited family of orbits with particularly useful parameters.

For example, the orbits closest to the Earth—known collectively as low Earth orbit (LEO)—are usually circular and less than 1000 km above the surface. This places them below the inner of the two Van Allen belts where radiation exposure is minimized. The only reason one would launch a spacecraft to an orbit within the Van Allen belts, which extend from about 2400 to 5600 km and from 13,000 to 19,000 km, would be to study the belts themselves. For all other spacecraft they are a nuisance.

Fig. 18   The planets of the solar system (the sizes of the planetary disks are approximately to scale).

[NASA]

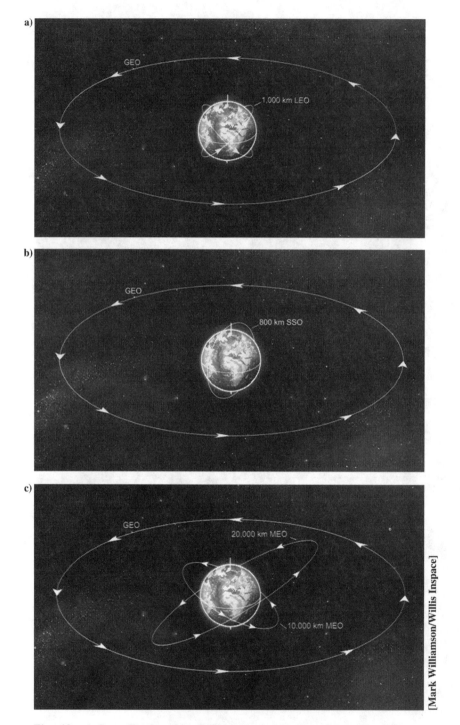

[Mark Williamson/Willis Inspace]

Fig. 19 a) Low Earth orbit (LEO); b) sun-synchronous orbit (SSO); and
c) medium Earth orbit (MEO) compared approximately-to-scale with
the altitude of geostationary orbit (GEO) at 36,000 km.

Spacecraft in LEO circle the Earth relatively quickly, typically with orbital periods of about 100 minutes. This makes these orbits good for satellites designed to image the Earth, because they are close to their subject and can cover large areas in a relatively short time. In fact, most imaging satellites are placed in a sun-synchronous/polar orbit (a LEO with a high inclination), which allows them to cover the entire Earth and revisit specific areas on a regular basis (see Fig. 19).

Most manned spaceflight activities have been confined to LEO because— apart from the Apollo Saturn V—no launch vehicles have been produced to place manned spacecraft in higher orbits. The Space Shuttle, for example, is designed to deliver payloads to LEO, and it is here that the ISS and most future manned space stations will reside.

The orbits beyond LEO are known as medium Earth orbits (MEOs) and lie at an altitude of around 10,000 km, placing them between the two Van Allen belts. Some MEOs are also known as intermediate circular orbits (ICOs), and, although ICO is often used as a synonym for MEO, the MEO classification is not restricted to circular orbits. This class of orbit is not as widely used as LEO because it offers fewer advantages, but it has been specified for a number of communications satellite constellations. Meanwhile, navigation satellite constellations such as the Global Positioning System (GPS) have been placed in higher MEOs above the upper radiation belt.

Most communications satellites are stationed in the higher-altitude geostationary orbit (GEO), which lies in the plane of the Earth's equator at an altitude of 35,786 km (22,237 miles). The orbit is sometimes abbreviated to GSO (for geostationary satellite orbit), or expanded to geostationary Earth orbit (which is tautologous because the Greek prefix "geo" refers to Earth). It is also known colloquially as Clarke orbit because the application of the orbit for satellites was first brought to public attention by the writer Arthur C. Clarke,[5] but this is not as useful as the descriptive term "geostationary." (The term geosynchronous is often heard in place of geostationary, but this refers to any orbit whose period of rotation is in some way synchronized with the Earth's rotational period; geostationary orbit is a special type of geosynchronous orbit.)

The key advantage of GEO is that satellites in that orbit appear stationary with respect to the Earth, which means that they can occupy a given orbital position related to the line of longitude above which they are stationed (see Fig. 20). This represents an extremely valuable resource for satellite communications operators and makes ground-based terminals much cheaper than those for equivalent LEO or MEO-based systems because they do not need to track the satellite. In addition, a system of three equally spaced satellites can provide full coverage of the Earth, except for the polar regions where the elevation angle of the receiving antenna is very low. As a result of its advantages, geostationary orbit has become the space-based equivalent of "prime real estate."

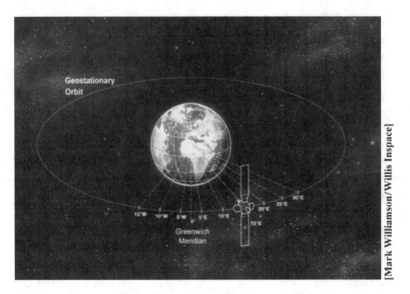

Fig. 20   Geostationary orbital positions are an important and valuable resource for the satellite communications industry.

Beyond LEO, MEO, and GEO, there are other so-called high Earth orbits or highly elliptical orbits (HEOs), some of which are used for specialized astronomy missions, for example. In addition, when coverage of high latitudes is important, satellites can be placed in high inclination, elliptical orbits, such as the Molniya orbit pioneered by the Soviet communications satellites of the same name.

By extension, most of these orbits are available around the other planets and major moons. In fact, spacecraft have already been placed in various orbits around Venus, Mars, Jupiter, Saturn, and the Earth's moon, and others are planned. One could easily imagine, for example, a future constellation of communications satellites in stationary orbit around Mars. (Following the Greek roots of geostationary orbit, it would be known as areostationary orbit after the Greek god of war, Ares.) Indeed, prior to its cancellation in 2005, NASA was planning to launch its Mars Telecommunications Orbiter in 2009.

Orbits around planetary bodies are not the limit, however. It is also possible to station spacecraft at a number of points in space where gravitational forces are balanced so that the spacecraft remains roughly stationary. They are known as the Lagrange or Lagrangian points after the Italian–French astronomer and mathematician Joseph Louis Lagrange who predicted that three astronomical bodies could move in a stable configuration, with one at each of the vertices of an equilateral triangle, if one of the bodies was small. [For any two celestial bodies orbiting about their common center of mass, five Lagrangian points (L_1–L_5), all in the same orbital plane, can be defined (see Fig. 21). L_1, the inner Lagrangian point, is, put simply, located between the two bodies where their gravitational attractions are balanced. L_2 and L_3, the outer Lagrangian points, are located at either end of a line passing through the two bodies (and L_1), where the combined gravity of the masses is balanced by centrifugal forces. L_4 and L_5, the equilateral points, are located on the orbital track of the smaller body around the larger, 60 degrees ahead and 60 degrees behind the smaller body.] The discovery of Jupiter's Trojan asteroids in the early 20th century proved the theory: they orbit in two groups, 60 degrees ahead and 60 degrees behind the planet, in its orbit about the Sun.

The L_1 Lagrange point between the Earth and the Sun, approximately 1.5 million km from Earth, is an ideal position for a solar observatory because it is never eclipsed by the Earth or any other planetary body. The first spacecraft to be stationed at L_1, in a so-called halo orbit, was NASA's International Sun-Earth Explorer 3 (ISEE 3), which, in 1978, became the first spacecraft to orbit a point in space rather than a celestial body. ESA's Solar and Heliospheric Observatory (SOHO), launched in 1995, is another successful example. Meanwhile, the L_2 point, on the opposite side of the Earth, has been suggested for deep-space observatory spacecraft.

For the Earth–Sun system, L_3 is on the opposite side of the Sun from Earth, L_4 is 60 degrees ahead of the Earth in its orbit, and L_5 is 60 degrees behind. L_4 and L_5 have been suggested as positions for space stations and space colonies.

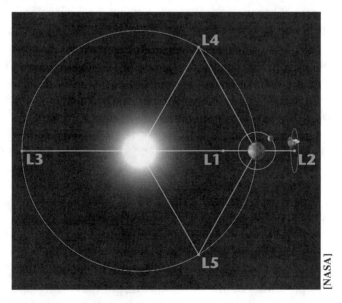

Fig. 21 Lagrange points for the Sun–Earth system, showing a spacecraft in a halo orbit at L_2 (image not to scale).

The message is clear: even without the surfaces of the planetary bodies, which present challenges of their own, there are many locations in space that are useful, some would say crucial, to mankind and modern society.

POINTS OF VIEW

So far, we have considered the space environment from a purely factual and descriptive standpoint: a "this is what there is to the best of our knowledge" approach. Beyond that, we should consider what the space environment means to different people, and what it does to their sense of curiosity, sense of wonder, or even their business sense.

The viewpoint of the scientist has already been alluded to. Astronomers, astrophysicists, and space scientists recognize the space environment chiefly as a subject of study and are concerned mainly with its existing properties. They will observe, study, catalog, and hypothesize about its many constituents and, if they have a soul, will wonder at the beauty of it all (Fig. 22).

As with all observational science, an important balance must be struck between the desire to discover more about the subject and the risk of influencing the properties of that subject by intervention. To cite a crude example, if scientists attempt to measure the temperature of a cool liquid with a warm thermometer, they will alter the very property they are trying to measure.

The balance in space exploration kicks in when remote and passive observation is not enough, for instance when it becomes necessary to collect rock samples to learn more about the composition of a planetary surface. At this point, the scientist has to consider to what extent the physical intervention of the spacecraft's, or indeed astronaut's, collection device will affect the purity of the sample. Digging directly underneath the descent engines of a lander, for example, would tend to produce a sample contaminated with engine exhaust products. Likewise, using a rover to extend the field of survey away from the lander could erase a subtle surface structure, pattern or coloring that was invisible from an imaging spacecraft in orbit.

On Earth, a field scientist can often identify an important geological structure or fossil before he raises his hammer, or spot a new species of crustacean before he puts his foot on it. Beyond the Earth, especially where exploration is limited to automated probes, things are not so easy. In both cases, the balance between the acquisition of knowledge and the alteration of the subject is important.

The viewpoint of the engineer, who designs the spacecraft, launch vehicles, and other hardware that enables space exploration, is somewhat different. Although many engineers are personally interested in the natural features and processes of the solar system, professionally speaking they are more interested in how the space environment relates to the spacecraft they design. It relates in two main ways: first, it provides a challenge to

[NASA]

Fig. 22   An astronomer's view of the space environment—something to study and observe.

their design skills, which they must use to overcome the factors that tend to damage or degrade their spacecraft; and second, it provides a useful set of locations—chiefly orbits and planetary surfaces—in which to station the spacecraft. To the engineer, then, the space environment is both an adversarial force and a useful resource (Fig. 23).

Among the challenges facing the engineer is the need to design reliable hardware capable of operating in the harsh thermal and radiation environment of space, often for long periods and usually without the possibility of direct human intervention. For example, satellites based in geostationary orbit, which is inaccessible to manned spacecraft, are typically designed to operate for up to 15 years.

An additional challenge, which is always present, is to produce the hardware within a tightly constrained budget and meet the delivery timescales demanded by the purchaser. For example, there would have been no point in delivering ESA's comet interceptor, Giotto, two years late: Comet Halley would have been and gone! Likewise, a commercial satellite operator

Fig. 23 An engineer's viewpoint—the space environment as a useful resource (in this case for the Galileo navigation satellite constellation).

is in business to make a profit; if his satellite is two years late reaching orbit, his profits and his shareholders will suffer.

The balance for the engineers is to design a product that can be built, launched, and operated as reliably as possible, while keeping costs and development timescales down to reasonable levels. As far as space—their basic resource—is concerned, they must also ensure that their products do as little damage as possible to that resource, ensuring, for example, that positions in geostationary orbit will be available for future satellites. The potential extent of that damage is the subject of the following chapter.

Although their roles are very different, space scientists and space engineers are two sides of the same coin. They have similar educational backgrounds, similar levels of knowledge of the space environment, and similar understanding of its importance. To those outside the fields of science and engineering, however, the space environment means something entirely different.

To *commercial satellite operators*, who form a majority among space users, it offers a place to station their satellites, which, if they have a good business plan, will earn them money. The space environment, specifically geostationary orbit, has provided this opportunity since the technology became available to place satellites in that orbit, in the 1960s. Since then, hundreds of satellites have been placed in GEO, and a great deal of money has been earned by selling the services they provide.

Other commercial users have come forward as space technology has developed. In the late 1990s, for example, the world of satellite-based

Earth imaging entered a new, commercial phase, using high-resolution sat-ellites in polar orbits. Advertisers have begun to use the International Space Station to promote their wares, by filming their adverts there, and the first fee-paying tourists have already started to arrive onboard. And for those not clever, lucky, or rich enough to visit space in their lifetimes, the first samples of cremated human remains have been launched into temporary orbits, later reentering the atmosphere in a fiery reenactment of their crema-tion. More and more, the space environment is becoming an extension of our terrestrial business and commercial environment.

To *government and military agencies*, the space environment is recog-nized as a place to station satellites for communications and electronic eaves-dropping, and as the ultimate "high ground" for reconnaissance and surveillance. Satellites are also used for missile detection and tracking, navi-gation for anything from aircraft carriers to cruise missiles, and for accurate position location of individuals. Navigation applications are also provided free-of-charge to civilians via GPS, operated by the U.S. Air Force.

Although space has not yet been officially designated as the fourth theater of war—after land, sea, and air—it is probably only a matter of time (Fig. 24). America's withdrawal from the antiballistic missile (ABM) treaty implies a renewed interest in the further militarization of space. In fact, construction of its new missile interceptor facility in Alaska began on 15 June 2002, the day the United States officially withdrew from the treaty.[6]

The wish of some, and the fear of many, is that low Earth orbit should one day house a fleet of missile-destroying spacecraft, developed under programs termed, successively, Strategic Defense Initiative (SDI)—otherwise known as "Star Wars"—and National Missile Defense (NMD). It remains to be seen whether ballistic missile defense will become a future use of the space environment.

Governments and their civilian agencies are also big users of space for communications, Earth observation, and so on. From their perspective, the space environment is an asset for weather forecasting, climate research, dis-aster monitoring, forestry management, and a host of science applications. In addition, their use of the space environment can also be a source of national pride, especially where manned spaceflight is involved. Moreover, it pro-vides employment for some of their citizens and acts as a wealth creator.

To the *industrial contractors* that supply the spacecraft for all of these applications, the space environment is effectively a source of business, contracts, and, ultimately, market share. Although the space industry might be a small segment of the aerospace industry, it remains an important part of many companies' portfolios and has great—some would say infinite—potential.

In addition to these institutional or corporate points of view, it must also be recognized that different cultures, generations, and individuals do not always

[EADS]

Fig. 24   A military viewpoint—space as a theatre of war.

view the space environment and its uses in the same way. Most western cultures see the potential of space exploration and development in economic or commercial terms, whereas many eastern cultures concentrate more on how space can benefit their underprivileged populations. On balance, the older generations (those who experienced the Apollo missions perhaps) are galvanized by exploration, while the younger generations show a sensitivity towards improving life on Earth.[7]

Indeed, our cultural view of space has also changed with time. In the early years of the Space Age, when launching even tiny satellites into orbit proved difficult, space was relatively unattainable and, like Mount Everest before it, was seen as something to conquer. The books published in that era reflect this in their titles, which include *Conquering the Sun's Empire*, *Assault on the Moon*, *War for the Moon*, and several examples of *The Conquest of Space*.[8–14] However, now that space has, in effect, been conquered, the cultural view as reflected in book titles is less adversarial.

VALUE?

The obvious conclusion is that the space environment means different things to different people. If the part of it used by one of the parties just discussed were suddenly to become unusable or unable to deliver the service they had come to expect, there would be commercial, scientific, military, or social repercussions.

This raises the question of value: if the space environment is so important to so many different groups of people and is used to such an extent in their daily activities, it must be valuable to them. If commercial operators wanted

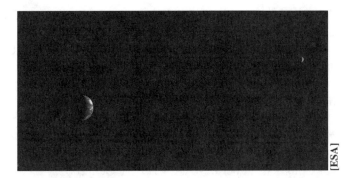

Fig. 25   A spacecraft's view of the space environment: Earth and Moon from Mars Express.

to sit down and calculate the contribution the space environment made to their businesses, they could assign a financial value—or asset value—to the part of the space environment they use. In fact, this has already been done in that satellites occupying individual geostationary orbital positions have been sold to other operators, thereby tacitly assigning the ownership of that position in space to the buyer. Although there is no official marketplace for orbital positions—they are assigned through an internationally agreed process by the International Telecommunication Union (ITU)—some of the haggling at ITU meetings rivals the best of bargain battles seen at market stalls.

The space environment obviously has a value for space scientists, because it offers a unique view of the universe and a unique environment—microgravity—in which to conduct experiments. It must also have a value to government and military agencies, not the least because they spend a huge amount of money dispatching their hardware and maintaining its operation in space.

So, if the space environment represents a valuable resource to so many different actors and agencies in society, it seems fair to assume that they would work to ensure that the resource remained available for their future use. Surely it is logical to assume that the resources of the space environment would be protected. In fact, neither assumption is correct.

The following chapter looks at how this apparent lack of care has affected the Earth's orbital environment.

REFERENCES

[1]Williamson, M., *The Cambridge Dictionary of Space Technology*, Cambridge Univ. Press, Cambridge, England, UK, 2001, p. 28.

[2]Fraser, R., *Once Around the Sun: the Story of the IGY 1957–58*, Hodder and Stoughton, London, 1956, p. 144.

[3]Stoiko, M., *Soviet Rocketry: The First Decade of Achievement*, David and Charles, Newton Abbot, England, 1970, p. 82.

[4]Martin, C.-N., *Satellites into Orbit*, George G. Harrap and Co. Ltd., London, 1967, p. 80.

[5]Clarke, A. C., "Extra-Terrestrial Relays," *Wireless World*, Oct. 1945, pp. 305–307.

[6]Barrett, R., "Construction of Missile Defense Installations Remains on Track," *Space News*, Vol. 14, Issue 1, 6 Jan. 2003, p. 4.

[7]Mendell, W., "Ethics and the Space Explorer," International Astronautical Congress, Paper 02-IAA.8.1.01, 1 Oct. 2002.

[8]Ordway, F. I., and Wakeford, R. C., *Conquering the Sun's Empire*, John Baker, London, 1965.

[9]Burgess, E., *Assault on the Moon*, Hodder and Stoughton, London, 1966.

[10]Caidin, M., *War for the Moon*, E. P. Dutton, New York, 1959.

[11]Lasser, D., *The Conquest of Space*, Hurst and Blackett, London, 1932.

[12]Ley, W., *The Conquest of Space*, Viking Press, New York, 1949.

[13]Ducrocq, A., *The Conquest of Space*, Putnam, London, 1961.

[14]Shelton, W. R., *Man's Conquest of Space*, National Geographic Society, Washington, DC, 1968.

Chapter 3

Earth's Orbital Environment

In the early years of the Space Age, mankind was concerned primarily with conquering space. Even getting there, let alone surviving there, was a challenge to his ingenuity and resolve. The process of placing a spacecraft in Earth orbit, or targeting a planetary body such as the Moon, was so difficult that almost any method was permissible in addressing that goal and there was little thought of any consequences that might arise from these actions.

Even when it had been proved—with Sputnik 1—that it was possible to place a satellite in orbit, it remained an extremely difficult and dangerous task. The statistics make discouraging reading: from October 1957, 12 of the first 20 attempts to orbit a satellite failed; and as of the 40th attempt, 23 had been abject failures. Amazing though it seems today, at the time of the first manned spaceflights about 50% of space missions were failing for one reason or another: either the rockets could not get their payloads into orbit, or the recovery systems could not get them down again. This made manned spaceflight an extremely risky business on both counts.

No one had the time or the inclination to consider the effect that hardware abandoned in orbit might have in the future. If the hardware had completed its job, or not worked in the first place, it was forgotten. As with many aspects of Earth-bound pollution, it has taken time to recognize the damaging effects of what we know now as "space debris."

Space Debris

It is important to understand what is meant by debris in the context of the space environment, because the terms "space debris," "orbital debris," and "space junk" are often used interchangeably and with little thought, by laymen and professionals alike.

Debris is, according to one dictionary definition, "fragments of something destroyed or broken" or "a collection of loose material derived from rocks . . . ," the word itself being derived from the 18th century French verb *debrisier*, to break into pieces.[1] As one might expect from a general dictionary of English, the definitions are generalized as far as possible and, when applied to space, appear to describe natural as opposed to man-made objects.

Although natural space objects such as meteoroids and micrometeoroids can be thought of as a form of space debris—in that they are effectively the detritus left over from the formation of the solar system, this does not reflect how usage of the term space debris is developing. These natural bodies are, in fact, a part of the space environment to the same extent as the atmosphere is part of the Earth's environment.

For the record, a micrometeoroid can be defined as "any small celestial body of natural origin, about the size of a grain of sand or smaller" [i.e., about 10 μm across (1 μm is a thousandth of a mm, or 10^{-6} m, and is also known as a micron.)]; a meteoroid is anything larger.[2] Once the body has collided with the Earth, or another planetary body, it becomes a meteorite or micrometeorite. (A meteor, by contrast, is a meteoroid made visible by frictional heating in a planet's atmosphere.)

Probably the most famous meteoroids are those that produce the annual Leonid meteor shower, so named because it appears to originate from the constellation of Leo. In fact, the Leonid meteors originate from the comet Tempel-Tuttle, which leaves a trail of ejected material in its path around the Sun. About once every 33.25 years, the Earth orbits through a particularly dense region of the trail, resulting in a rise in the number of meteors observed. Although the most recent peak occurred in November 1999, the Leonids received their greatest publicity the previous November, when it was believed that satellites orbiting the Earth were at risk. However, despite the fact that there were many more satellites in orbit than during the previous peak, in 1966, there were no reports of suspected Leonid damage (Flury, W., personal communication, 25 May 2004).

The point is that these and other natural objects constitute a background flux of material that we can only guard against; we cannot turn it off, and we cannot decrease its flow (Fig. 26). In a worst-case scenario we could increase it, perhaps by destroying an asteroid or causing a large explosion on a moon, but its development is largely out of our hands. However, what has come to be known as space debris is quite different.

Space debris is a blanket term for any man-made artifact discarded, or accidentally produced, in space, either in orbit around a planetary body (when it is also known as orbital debris) or on a trajectory between planetary bodies.[3] In contrast to the coverage of the standard English dictionary definition just quoted, space debris is not confined to fragments of something destroyed; it can also include abandoned items that are no longer useful. As such, space debris includes satellites that have reached the end of their lives, spent launch vehicle stages, hardware accidentally released by astronauts (including bolts, lens caps, and the almost legendary astronaut's glove that floated out of the hatch of a Gemini capsule), as well as the remnants of spacecraft and launch vehicles that have exploded or been hit by other debris.

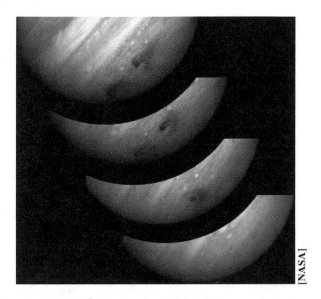

[NASA]

Fig. 26   The result of a natural debris impact on Jupiter (Comet Shoemaker-Levy 9 in 1994).

Having said this, the term "space debris" can have different meanings, depending on the organization or individual using it. In the United States, for example, the term is often used to describe both natural and man-made debris, with the subset "orbital debris" reserved for man-made items in orbit. A useful alternative is to refer to man-made debris as "artificial space debris." In the final analysis, it does not matter which definitions are used as long as a common understanding is reached.

SOURCES

In principle, anything we send into space can end up as debris, but most of the debris comes from one of two main sources: launch systems and satellites.

Because of the inherent difficulties involved in launching objects into space, it is common to adopt the simplest possible approach. During the early years of the Space Age, this usually meant stacking a number of rocket stages on top of each other, strapping on the occasional solid rocket booster, and bolting the satellite on top. Launch vehicle stages and the satellites themselves were separated by firing explosive charges, which severed the bolts or metal bands that held the segments together during launch.

Although debris from the lower-stage separations had insufficient velocity to enter orbit, that from upper-stage and satellite separations typically accompanied the useful hardware into orbit. This meant that each launch created a cloud of shattered clamp bands, connectors, and associated debris with the same forward velocity as the satellites, and some with a lateral velocity component, which spread it into other orbital paths. Some of the material attained orbit temporarily and then reentered as a result of atmospheric friction, whereas other bits remain in orbit to this day.

Moreover, for some upper stages that was not the end of the story. Those that attained sufficient velocity to enter a long-term LEO usually retained in their tanks an amount of residual fuel and fuel vapor, surplus to that required to place the payloads into orbit. As a result of the heating effect of the Sun, this fuel sometimes produced an explosion, contaminating a range of orbits with an expanding cloud of debris.

In technical circles, such debris-producing incidents are known as fragmentation events, although the term includes both energetic events, or explosions, and less violent events, which simply cause an object to break into fragments.

As it turns out, explosions—mostly of rocket stages—are the biggest source of space debris.[4,5] According to NASA's *History of On-Orbit Satellite Fragmentations*, published in 2001, the debris from these explosions accounted for nearly 40% of all cataloged objects *remaining* in orbit.[6] Moreover, of the more than 4150 space missions flown since 1957, just 10 were responsible for 21% of all of the orbiting debris, as cataloged in the official U.S. Satellite Catalog, which lists objects in orbit and those that have decayed or been boosted into interplanetary space.

The first in-orbit explosion occurred quite early in the Space Age, on 29 June 1961, when the Able Star upper stage used to launch the Transit 4A satellite exploded. It produced 296 cataloged pieces of debris, 191 of which

[NASA]

Fig. 27   Although the Space Shuttle's external tank does not reach orbit, those launch vehicle upper stages that do can become a source of debris.

were still in orbit in 2001. Since this first event, more than 170 fragmentations have been identified, often occurring several years after the launch. For example, an upper stage of the Delta vehicle that launched the Nimbus 6 weather satellite in June 1975 exploded in May 1991, producing 237 cataloged debris pieces.[6]

Although it is often said that time is an element in the fragmentation process, largely because of the continual movement of a low-orbiting object between sun and shadow and the consequent heating and cooling cycles, this is not always the case. On 4 October 1990, for example, a Chinese Long March 4A final stage, which had been in orbit for only a month, unexpectedly exploded into 83 trackable objects. Others have broken up within days or even hours of reaching orbit.

Moreover, despite the introduction of new technology, it is evident that upper-stage explosions are neither confined to the past or to the world's less experienced space nations, as shown by incidents with two American launch vehicles:

1) On 7 February 1994, the Titan 2 second stage used to launch the American Clementine spacecraft the previous month produced some 365 trackable pieces of debris in LEO. Their orbits were short lived because of the low altitude (200 km), but because a Shuttle mission (STS-60) was underway at the time special attention was paid to the incident.

2) On 3 June 1996, the Pegasus upper stage—the Hydrazine Auxiliary Propulsion System (HAPS)—used for the MSTI 2 mission in May 1994 exploded into more than 750 trackable debris objects. Their average altitude of 625 km was only 25 km above the orbit of the Hubble Space Telescope.

With regard to the timing of these events, it is notable that the offending stages were launched only four months apart, but suffered explosions more than two years apart; they were like timebombs with unpredictable fuses.

The rocket stages just discussed used liquid propellants, but the remains of solid-propellant rockets can also produce space debris, albeit of a special kind. For many years, solid rocket motors have been used—in their hundreds—to boost satellites into orbit and other spacecraft onto interplanetary trajectories. A common method of enhancing the performance of these motors involves adding finely ground metal particles, aluminum for example, to the solid fuel mix. Unfortunately, in operation, a single motor can produce a wake of some 10^{20} (100 million million million) particles, ranging in size from 0.0001 to 0.01 mm, and although they are thought to reenter fairly quickly, the fraction that remains exceeds the natural micrometeoroid flux.[7]

In addition to this high velocity flux, around 1% of propellant mass leaves the motor at low velocity towards the end of the burn: it ranges in size from a few millimeters to 2 or 3 cm and is known as rocket motor "slag" (Flury, W.,

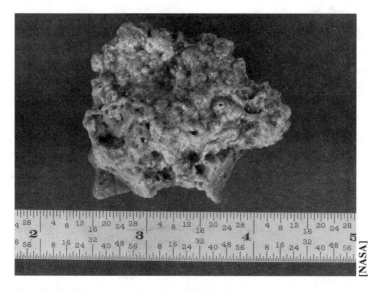

Fig. 28   Solid rocket motor "slag." This piece of aluminium oxide slag
was recovered from a test firing of a Shuttle solid rocket booster.

personal communication, 25 May 2004), see Fig. 28. In contrast, although it
depends on the propellant and the manner of its release, most liquid propellants
released into space will atomize, or form solids that will later sublimate; the
same is true for water (Johnson, N., personal communication, 25 March 2004).

Now whereas anyone can appreciate the effect of being hit by a rocket
stage or propellant tank, it is more difficult to see how particles a fraction
of a millimeter across could harm a spacecraft. The problem, however, is
not so much the objects' size and mass, but their velocity, which can give
them the energy of a high-power rifle shell. Because an object's kinetic
energy is given by the expression $\frac{1}{2} mv^2$, where m is the mass and v the velocity, the square of the velocity has a huge effect on the energy.

For example, the velocity required to escape the Earth's gravity—a
vehicle's escape velocity—is 11.18 km/s (over 30 times the speed of
sound); this gives an indication of the magnitudes that need to be considered.
In fact, the average collision velocity between two cataloged objects in LEO
is about 10 km/s (Johnson, N., personal communication, 25 March 2004). A
simple calculation shows that a 1-g particle moving at this speed has the
same energy as a 1-tonne car traveling at 10 m/s (about 36 km/h).

If an object moving at 10 km/s in one direction meets another moving in
the opposite direction, their relative velocity (and impact energy) is of course
even greater. Indeed, average meteoroid velocities are thought to be between
20 and 40 km/s (Johnson, N., personal communication, 25 March 2004),
whereas for the Leonids, typical particle velocities of around 70 km/s are
believed to be common.[6]

Although the expected maximum collision velocity in GEO is understood to be only a fraction of that in LEO—around 800 m/s[9]—this is quite enough if your satellite is on the receiving end! Luckily, the probability of impact for a satellite in GEO has been estimated to lie between 10^{-6} and 10^{-5} in a given year (a one-in-a-million chance).

The most obvious effects of impacts include damage to spacecraft structures, the destruction of equipment on unmanned spacecraft, and the potential depressurization of a manned module. Less obvious perhaps is the effect they can have on the attitude, or pointing direction, of a spacecraft. Earth-orbiting spacecraft have a finely balanced existence, which can be disturbed by extremely small forces, including the gravitational pull of the Sun and the Moon and even solar radiation pressure. The impact of a physical object can unbalance a satellite to the point where its antennas, sensors, or cameras no longer point towards the Earth or an object of astronomical interest. Although spacecraft incorporate automatic systems to recover their pointing, an impact is at the very least disruptive; at worst, it can threaten the mission (Fig. 29).

In addition, the continual impact of very fine debris can lead to the gradual degradation of spacecraft surface coatings, such as thermal insulation and solar reflectors, which can cause a spacecraft's temperature to rise, whereas the degradation of solar arrays can reduce their ability to generate power.

Even less obvious than individual impacts and continual "sandblasting" are electrical charging effects, which can occur if the incident particles are charged. A buildup of charge on a spacecraft usually leads to electrostatic discharges, which can damage electronic components—for example, by inducing short circuits—and can even cause the loss of the spacecraft. Indeed, ESA's Olympus communications satellite is believed to have failed because of an impact from the 1993 Perseid meteor cloud (formed from the remnants of Comet Swift-Tuttle). Apparently, an impact on a solar array generated an electrical charge, which was channeled into the spacecraft where it overloaded an electrical circuit.

Given the amount of man-made debris in orbit, it seems logical to conclude that, under slightly different circumstances, the object could just as easily have come from another spacecraft. In fact, many satellites have produced debris in the course of their normal operation, mainly during the deployment of items of hardware such as solar arrays and antennas. As a result, pyrotechnic bolts and clamp bands on satellites, similar to those employed as stage separation devices, have added to the debris population over the years.

Since the Able Star explosion in 1961, about three dozen satellites have produced debris in nonexplosive events. Although most of these events have occurred in low Earth orbits (because that is where the majority of satellites have been placed), they are not unknown elsewhere. The first known fragmentation event in GEO—the most popular orbit for communications satellites—was the explosion of a battery on the Russian Ekran 2 satellite,

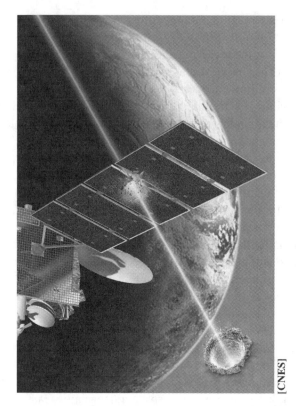

Fig. 29   Impacts from debris—natural or man-made—can damage satellites.

which occurred on 25 June 1978. Significantly, the event was not detected by the American authorities, but was eventually disclosed by the Russian government in an orbital debris meeting held in February 1992.[6]

On top of these "accidental" events, at least 10 known satellite fragmentations were deliberate acts. They include the intentional detonations of nine Soviet interceptor craft in tests of its Fractional Orbital Bombardment Satellite (FOBS) system, developed in the 1970s as a reaction to concerns regarding the proliferation of U.S. reconnaissance satellites. The satellites were designed to approach a target satellite and explode, creating a cloud of shrapnel that would destroy the target. The other example concerned a U.S. Air Force antisatellite (ASAT) test of September 1985, which blasted the 850-kg Solwind satellite into more than 285 tracked pieces. In this case the projectile was launched from an F-16 fighter aircraft, and 95% of the debris was understood to have been removed from orbit by atmospheric friction by early 1990.[10]

Arguably the *most deliberate* act of introducing debris to LEO was the release of a cloud of copper "needles" as part of a U.S. Air Force communications program known as Project West Ford. Its intention was to girdle the

Earth with a passive reflector formed from 18 mm lengths of copper wire, electrical dipoles from which digitized speech signals could be reflected. The experiment was first attempted on 21 October 1961 when the Midas 4 satellite, in a polar orbit, ejected a container containing 350 million dipoles. They were embedded in naphthalene, which was supposed to evaporate, slowly releasing them to form an orbital belt some 8 km wide and 40 km deep. Ground radar subsequently located the container but failed to detect the dipoles, and it is assumed that they did not disperse.[11]

In May 1963, however, a 23-kg payload of dipoles was ejected from the Midas 6 satellite at an altitude of about 3600 km. Although the experiment was deemed a success at the time, the long-term consequences of a malfunction in the dipole dispensing system has cast a shadow on the endeavor. Although individual dipoles, correctly dispensed, were expected to reenter the atmosphere relatively quickly as a result of solar radiation pressure, the malfunction is believed to have formed clumps that remain in orbit, as a potentially destructive force, to this day (Johnson, N., personal communication, 25 March 2004).

One final example of pollution in space implies that Project West Ford was simply a sign of the times. According to Dean Rusk, who was secretary of state under U.S. Presidents Kennedy and Johnson, before the Nuclear Test Ban Treaty was invoked in 1963, President Kennedy received a request from the military for permission to detonate a nuclear weapon on the inner edge of the Van Allen belts.[12] Asked whether this would have any detrimental effect on the belts, Kennedy's scientific advisors initially reported that it would not and the test went ahead. However, says Rusk, two days after the test, the same advisors came back and apologized that they were "in error by a factor of two thousand." The explosion created "a little Van Allen Belt of its own," and a lesson was learned.

TRACKING

How do we know about the debris in orbit? One of the reasons it took so many years for the space community to come to terms with the debris-producing potential of its own products was the lack of evidence. Once hardware had been launched into space, it was, to a large extent, out of sight and out of reach.

Given the difficulties involved in returning space hardware to Earth, engineers had to devise a means of remote observation of orbital debris, largely using ground-based radars. The leading authority for debris tracking is the U.S. Space Surveillance Network (SSN), which publishes the U.S. Satellite Catalog. The SSN was originally operated by the North American Aerospace Defense Command (NORAD) and from 1986 by U.S. Space Command, whose mission was later absorbed by U.S. Strategic Command.

Other facilities, which track by radar and optical means, have been installed by other spacefaring nations, but only the United States and Russia have dedicated space surveillance systems, that is systems capable of providing an overview of all orbiting objects larger than a given minimum size (Fig. 30). The SSN, for example, tracks objects in LEO at least 10 cm in diameter on a routine basis.

Nevertheless, Europe has two world-class radar facilities for spot checks (as opposed to systematic observations) and the intellectual resource of its own Space Debris Advisory Group (SDAG). Both the German Research Establishment for Applied Science (FGAN) tracking and imaging radar (TIRA), at Wachtberg near Bonn, and the French military ship *Monge* can detect objects of about 2 cm in size at a range of 1000 km. Other tracking and imaging facilities are located in Scandinavia. China has a single tracking

Fig. 30   Debris tracking facilities: a) 70 m antenna at Goldstone, California; b) Air Force Maui Optical and Supercomputing (AMOS) site in Hawaii; c) Haystack and HAX radars operated by MIT's Lincoln Laboratory.

station at the Purple Mountain Observatory in Nanjing, which opened in 2005, while France is developing an experimental system, called GRAVES, but with significantly less capability than the SSN (Flury, W., personal communication, 25 May 2004).

Radar systems with very large antennas, such as the Massachusetts Institute of Technology's (MIT) Lincoln Laboratory's Haystack Long Range Imaging Radar, which has a 36-m dish and a 400-kW, 10-GHz transmitter, are capable of detecting objects down to 1 cm at an altitude of 1000 km. It was the similarly impressive capability of the 70-m-diameter Goldstone radio telescope that allowed the detection of liquid-sodium/potassium coolant leaking from one of Russia's retired nuclear-powered radar ocean reconnaissance satellites, or RORSATs. Some 70,000 droplets were found in near circular orbits at 700–1000 km, to which the reactors were designed to be boosted before the rest of the satellite reentered.[7] The size of the droplets ranges from 6 mm to 4.5 cm, and their total mass is estimated as anywhere between 50 kg[13] and 165 kg.[14] It is believed that some two dozen surviving nuclear reactors from the Cosmos series, together with other Soviet or American debris, account for an estimated one tonne of uranium 235 and fission products currently in orbit.[7]

Following a detailed analysis of impact features on returned hardware, it has been estimated that there is also about a tonne of orbital debris fragments smaller than 1 cm and 300 kg smaller than 1 mm. Considering that the estimated total mass of active and inactive man-made objects within 2000 km of the Earth's surface is only about 2000 tonnes, the mass of debris is significant.

Radar techniques have shown, unsurprisingly, that space debris is concentrated in the orbits most commonly used by spacecraft, namely the low Earth orbits (Fig. 31). Most orbital debris resides within 2000 km of the Earth's surface, although the amount varies significantly with altitude, with concentrations around 800, 1000, and 1500 km, reflecting the most popular orbital heights.

Self-evidently, the further the objects are from Earth, the more difficult they are to detect, and so a debris concentration in LEO might be more perception than fact. This reduction of radar sensitivity with altitude is crucial for the satellites in geostationary orbit, because the smaller, but no less dangerous, debris objects remain undetectable. Moreover, the ability to detect objects depends on many factors, including shape, material composition, and the eccentricity of their orbit. Thus, as a general rule, radar systems tend to be used for debris in LEO, whereas optical systems are used for higher-altitude orbits, typically for detecting debris as small as 20 cm in the vicinity of GEO (Johnson, N., personal communication, 25 March 2004). In fact, in 2000, ESA's 1-m Zeiss telescope at the Teide Observatory in Tenerife detected and determined the orbits of geostationary debris as small as 10–15 cm (Flury, W., personal communication, 25 May 2004).

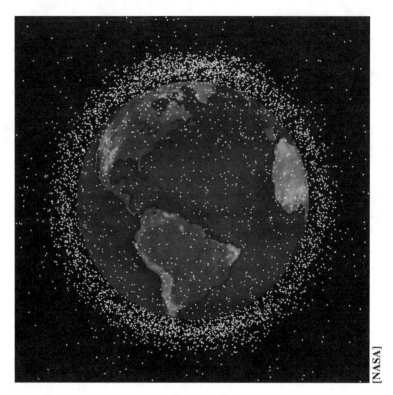

[NASA]

Fig. 31   Computer generated graphic of tracked debris objects in low Earth orbit (LEO).

One of the most recently inaugurated electro-optical systems is the Okno ("window") facility in the former Soviet republic of Tajikistan. Operated by the Russian Space Forces, it is reportedly designed to observe objects in orbits as high as 40,000 km, which encompasses geostationary orbit at 36,000 km and the apogee of the elliptical Molniya orbit. Its location, in a mountain region 2200 m above sea level, provides relatively clear skies and a stable atmosphere for optical transmission.[15] Among plans for future systems is the U.S. Air Force's proposal for the Satellite Active Imaging Testbed (SAINT), which would use lasers to illuminate satellites in LEO and produce images of 5-cm resolution or better.[16]

Although the size and mass of the debris objects are obviously important, so is their number, because larger numbers imply wider distribution (see Table 2 and Figs. 32–34). By January 2005, the official U.S. Satellite Catalog contained almost 28,500 objects, noted since the launch of Sputnik 1 on 4 October 1957. More than 9200 objects were still in orbit around the Earth, some 6300 of them in LEO and more than 1000 in the vicinity of GEO. The figures include both operational and defunct satellites, abandoned rocket bodies, and other categories of space debris.

Table 2 Debris objects in LEO and other orbits by number, mass, and area (as of January 2005)

Quantity	LEO	Non-LEO	Totals
Number	6344	2707	9051
Mass, kg	2.24×10^6	2.76×10^6	5.00×10^6
Area, m^2	1.49×10^4	2.66×10^4	4.15×10^4

As of 30 March 2005, there were 9316 tracked objects in total, of which 2935 were defined as payloads and 6381 rocket bodies and debris. Coincidentally, the United States and Russia have contributed to the orbital debris environment in roughly equal measures, with approximately 4000 objects each.[17] Although the total number of tracked objects increased by several hundred between 2002 and 2005, it is difficult to say whether the number of objects has actually increased, or simply whether the number detected has increased. It is, however, clear that there are a large number of readily detectable objects in orbit.

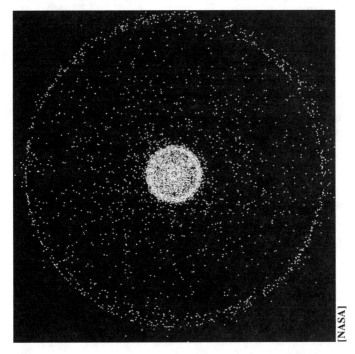

Fig. 32 Computer generated graphic of tracked debris objects from low Earth orbit (LEO) to geostationary orbit (GEO), viewed from above the north pole.

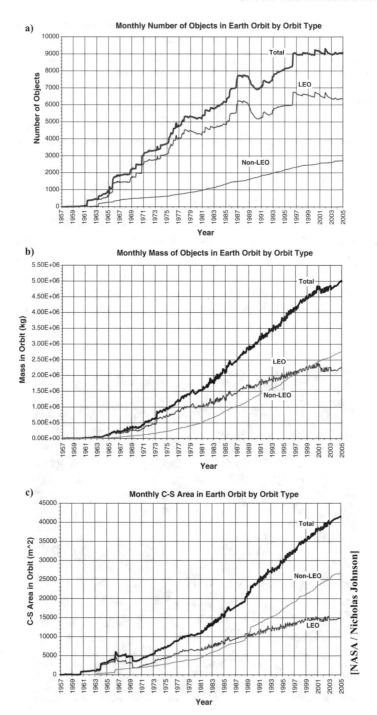

Fig. 33   Space debris tracked in Earth orbit (by *orbit* type): a) number of objects; b) mass of objects; c) cross-sectional area of objects.

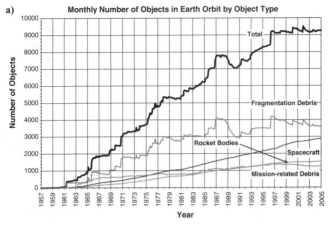

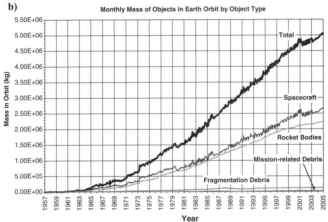

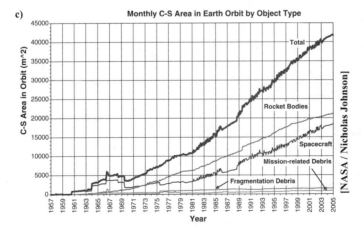

Fig. 34 Space debris tracked in Earth orbit (by *object* type): a) number of objects; b) mass of objects; c) cross-sectional area of objects.

According to NASA estimates, the population of particles between 1 and 10 cm in diameter is greater than 100,000, whereas the number of particles smaller than 1 cm probably exceeds tens of millions. Individual objects as small as 3 mm have been detected by ground-based radars, providing a basis for a statistical estimate of their numbers, while assessments of the debris population smaller than 1 mm can be made by examining impact features on the surfaces of spacecraft returned from orbit by the Space Shuttle. Though useful, this is limited by the Shuttle's operational capabilities, which confine it to altitudes below 600 km. What this means, in the final analysis, is that the orbital debris population is likely to be greater than we can either observe or estimate.

Interestingly, the proliferation of orbital debris is also of concern to Earth-bound astronomers because long-duration exposures of astronomical objects are increasingly degraded by debris trails. Because an average of between two and three trails per plate is recorded by the major Schmidt telescopes,[18] the astronomical community needs no more proof of the existence of space debris. Nevertheless, evidence held in one's hands is worth a library of images and radar traces.

PHYSICAL EVIDENCE

The relative lack of access to space—a function of the technical difficulty and the expense—has been a bugbear for the space community since the dawn of the Space Age. Most of the hardware we launch into space is either nonreturnable or inaccessible, so that we have no direct way of assessing its condition or performance following launch. Were it not for sophisticated radio telemetry—a technological field brought to maturity by the requirements of distant spacecraft—we would have little knowledge of spacecraft status.

Although this situation will continue for the foreseeable future for orbits and trajectories far from Earth, the hardware returned from LEO by the Space Shuttle has provided indisputable proof that space debris causes damage to spacecraft.

The first useful data were obtained from portions of the Solar Maximum Mission (SMM) satellite (Fig. 35), which was serviced by a Shuttle crew in April 1984, but more important was that derived from a spacecraft deployed on the same mission, the Long Duration Exposure Facility (LDEF). LDEF was designed to monitor and record the effects of the space environment and, with the help of the Shuttle, deliver its results directly to scientists and engineers on Earth.

Recovered after five and a half years in orbit, LDEF bore the scars of its mission in LEO (see Fig. 36). According to the recovery Shuttle's commander, the damage to thermal blankets on the space-pointing end of the satellite

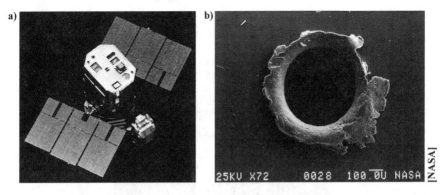

Fig. 35   a) Solar Max in orbit (note astronaut in Manned Maneuvering Unit at lower right); b) a hole made by orbital debris in a panel on Solar Max.

made it look like a "rolled-up sardine can." Some experiment components were missing while others were "barely hanging on," a solar-cell module was unattached and floating nearby, and a mass of small particles circled the satellite's midriff, glistening in the sunlight.

Some of the damage was caused by micrometeoroids, whereas other degradation was put down to radiation and thermal effects. Around 34,000 micrometer-sized impacts were counted, giving LDEF's panels the appearance of shotgun targets, along with larger craters up to 5.25 mm in diameter.[18] Although the most severe damage was found, as expected, on the leading surfaces, the number of impacts on trailing surfaces was significantly higher than predicted, indicating the presence of debris in intersecting orbits. Interestingly, a timing experiment showed that impacts sometimes came in 3–5-minute bursts, which supports the existence of concentrations or clouds of debris particles.[10]

In December 1993, a Shuttle returned a solar array from the Hubble Space Telescope (Fig. 37), which at that time had spent three and a half years in LEO. Because the array was designed and manufactured in Europe, its condition was of interest to the European Space Technology Centre (ESTEC), which undertook an analysis of the array. It showed that most impacts came from particles sized from 0.1 μm to a few millimeters in diameter.[19] The largest impact found was about 7 mm in diameter, while another impact had chipped off the corner of a solar cell. There were about 150 complete penetrations, but none was believed to have compromised the function of the array because impacts are usually only a problem if they sever a connection.

On the other hand, ESA's Eureca spacecraft, which was delivered to LEO in 1992 and returned the following year, provided evidence that impact-induced discharges can damage solar arrays. Analysis showed a

a)

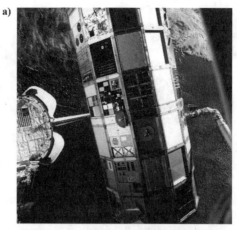

b)

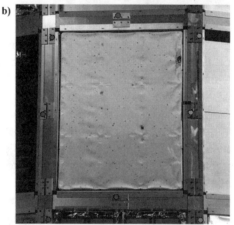

c)

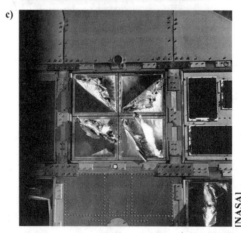

[NASA]

Fig. 36   a) Long Duration Exposure Facility
(LDEF) retrieval from orbit by Space Shuttle;
b) close-up view of panel at top center in (a);
c) peeling metal foil on LDEF experiment.

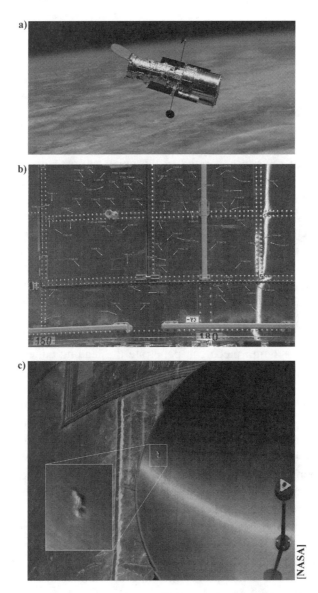

Fig. 37 a) Hubble Space Telescope in orbit; b) analysis
of the spacecraft body revealed multiple debris hits; c)
one impact penetrated an antenna reflector.

solar-array burn, measuring about 7 by 8 cm, where a discharge had caused
damage to several adjacent cells.

But most, if not all, of this damage could be blamed on the background
flux of natural debris. Where is the physical evidence of man-made space
debris?

Some of the most compelling evidence comes from the Shuttle itself, which has suffered damage from the tiniest samples of space debris, believed to have originated from orbiting satellites. In June 1983, for example, a window of the Challenger orbiter was cratered by an object that chemical analysis later revealed to be a particle of paint (see Fig. 38). The crater contained titanium oxide with traces of carbon, aluminium, potassium, and zinc, consistent with the composition of thermal paint containing an organic binder.[20] It has since been hypothesised that atomic oxygen—single atoms of oxygen generated in the upper atmosphere by solar ultraviolet radiation—can erode the organic binders in spacecraft paint, causing it to flake and break off. Calculations showed that a 0.2-mm paint fleck, traveling at a relative velocity of between 3 and 6 km/s, produced the 4-mm-diam crater.[8] One needs little imagination to estimate the result of a larger impact.

In fact, this was far from an isolated incident. In March 1997 it was reported that, in the previous 16 months, space debris had caused damage to 13 Space Shuttle windows sufficient to require their replacement (at a cost of some $40,000 each). Under the right conditions, a piece of debris as small as 0.4 mm can cause enough damage to require a window's replacement, whereas a 1-mm piece can penetrate the leading edge of a Shuttle's wing, which, following damage sustained during reentry, would need repairing.[21] Although the destruction of the Columbia orbiter in January 2003 was the result of an impact from external tank insulation detached during launch, rather than orbital debris, it provided a stark demonstration of the effects of wing leading-edge damage.

In fact, it has been known for some years that the brittle reinforced carbon-carbon (RCC) composite used for the leading-edge thermal protection system is particularly prone to damage. Although it represents only 3% of the total orbiter surface, the RCC accounts for approximately 40% of the risk to the vehicle from meteoroid and orbital debris impact.[22]

Indeed, the risk to a Space Shuttle orbiter is significant. For example, the risk of "critical penetration," meaning a potential loss of vehicle and crew, is about 1 in 250 for a typical mission to the International Space Station (ISS), which makes orbital debris the single largest threat to Shuttle operations from launch to landing (Johnson, N., personal communication, 25 March 2004). Little wonder then that, as a result of the orbital debris impacts, the standard flying configuration of the Shuttle since the mid-1990s has been tail first and upside down (with its payload bay facing the Earth), to help protect the most fragile components and reduce the potential for disaster.

Although somewhat less conclusive than the spacecraft paint example, some of the first measurements of man-made space debris were taken in

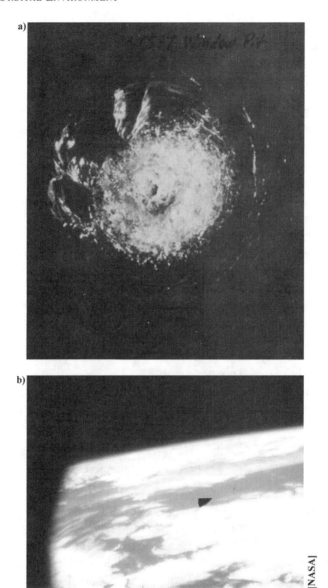

Fig. 38 Space Shuttle as victim and perpetrator: a) debris crater in shuttle window on STS-7 mission; b) tile fragment from Columbia's own thermal protection system photographed in January 1986.

1973–1974 by the Cosmic Dust Experiment on the Skylab space station: analysis of residues in impact craters showed "a high incidence" of aluminum. In addition, the windows of the Skylab 4 command module, analyzed by scanning electron microscopy, showed that about half the pits were lined with aluminum. Their size distribution and frequency matched what one would expect from solid rocket motor firings in space.[20]

Impacts have also been recorded on the Russian space stations. For example, in July 1983, a month after the first Shuttle window impact, an object believed to be a micrometeoroid hit the window of a Salyut space station. Likewise, the Mir station suffered its fair share of impacts from small orbital debris and meteoroids. In May 1988, for instance, a high-velocity particle created a window crater and fracture lines covering a width of some 6 to 8 mm.[10] Then, in September 1993, following heavy bombardment from the Perseids, an extra vehicular activity (EVA) revealed a 10-cm-diameter hole in one of the solar arrays and 65 much smaller impact craters.[23] Surprisingly, the capability of the array did not appear to have been significantly affected.

As indicated, the detection of relatively small debris objects has been a largely *ad hoc* process, and there has been no concerted effort to characterize the population by direct detection. This is strange considering that America's very first satellite, Explorer 1, carried, as part of its minimal payload, a microphone designed to detect micrometeoroid impacts. Although the satellite remained operational for less than four months, it detected an average of one impact every 15 days. The same experiment repeated today would almost certainly record a greater number.

In the final analysis, despite the intermittent data, it seems we must conclude that all spacecraft—particularly those in LEO—are being hit on a daily basis by very small meteoroids and/or orbital debris.

SPACECRAFT COLLISIONS

So much for small pieces of debris striking individual spacecraft, but what about a collision between two spacecraft or rocket stages? To what extent does this common fictional scenario adhere to reality?

Although the chances of a collision are currently relatively small, they are not insignificant, as shown by an event on 24 July 1996, when the French microsatellite Cerise was hit by debris from an Ariane rocket's third stage, which had exploded in November 1986, generating over 700 fist-sized debris fragments.[24] The relative speed at the time of collision was about 14 km/s or 50,000 km/h.[10] It severed the satellite's gravity-gradient stabilization boom (Fig. 39), which, along with the rest of the satellite and the offending debris object, was still being tracked in 1997.[10] The boom finally reentered in 2000 (Johnson, N., personal communication, 25 March 2004).

Despite the impact, the satellite and its payloads continued to work normally following recovery operations and reprogramming of the onboard computers. It was even possible, using attitude dynamics analysis

Fig. 39   Artist's impression of debris from an Ariane third stage striking the Cerise microsatellite in 1996.

techniques, to conclude that the 6-m-long boom had been reduced to about 3 m.[25] Although one normally considers collision probabilities to increase with cross-sectional area, for obvious reasons, it is interesting to ponder that, in this case, a piece of orbital debris struck a satellite structure no more than a few centimeters in diameter.

Significantly, this was the first *validated* collision between two *cataloged* objects, but this is far from saying it was the first spacecraft-debris collision. Because it is impossible to track and monitor all debris objects, it is impossible to say how many might have come into contact.

A more recent, validated collision occurred on 17 January 2005, when a U.S. Thor Burner 2A upper stage (used to launch a satellite in 1974) and a fragment of the third stage of a Chinese LM4 launch vehicle (which exploded in March 2000) collided some 885 km above the Earth. Data from the SSN indicated that three additional debris objects—large enough to be detected and cataloged—were released from the U.S. rocket body. The episode was reported in the April issue of *The Orbital Debris Quarterly News*, a publication of the Johnson Space Center's Orbital Debris Program Office, which also mentioned an earlier collision that had only recently come to light as a result of examining historical tracking data. It concerned a collision in December 1991 between a defunct Russian navigation satellite, Cosmos 1934, and debris from Cosmos 926. Additional debris was created in this case too, but the fragments were too small to be tracked.[26]

Luckily the chances of a collision between two *active* spacecraft are small because space is so big: one need only consider the volume of space contained within a sphere measured on the 1000-km scale of orbits. In geostationary orbit, for example, each one-degree of orbital space is equivalent to about 735 km, so that even if satellites were stationed one-tenth of a degree apart, they would, on average, be separated by 73.5 km of space. In reality, most are much further apart.

Despite this, the experience of the French space agency, Centre National d'Etudes Spatiales (CNES), has shown that even small numbers of spacecraft in similar orbits can be of risk to each other if one of them becomes uncontrollable. For example, the loss of the SPOT-3 Earth observation satellite in November 1996 forced (CNES) to study collision risks in some depth because of the possibility that it might collide with other SPOT satellites in similar, sun-synchronous orbits.[27]

The Agency's monitoring program led to an avoidance maneuver in 1997, when SPOT-2 was moved to avoid, not its sister spacecraft, but debris from a Thor Agena rocket stage believed to have come within 400 m of the spacecraft. The maneuver meant that SPOT-2 was unable to take images for two days, which had obvious commercial repercussions. Then, in May 1998, the Russian Meteor I-16 satellite came within 91 m of the spacecraft: in this case, there was insufficient time to make an avoidance maneuver but, thankfully, no collision occurred.

Of course, there is always the possibility of collision between two spacecraft intended to dock with each other, as the collision in 1997 between a Progress supply craft and the Mir space station showed only too well. Among other things, it damaged one of the station's solar-array panels. Likewise, the collision, in 2005, between NASA's Demonstration of Autonomous Rendezvous Technology (DART) spacecraft and its target, a retired U.S. military satellite, indicated the potential for debris creation (Fig. 40). Taken together, these incidents show that rendezvous and docking techniques—both remotely piloted and autonomous—have still to be perfected.

Currently, in-orbit docking attempts take place on a more or less monthly basis. If we are to contemplate a future in which regular visits are made to space stations, and autonomous refueling and deorbiting spacecraft dock with satellites on a frequent basis, the potential for further debris creation must be addressed.

SOLUTIONS

Given the amount of debris in orbit, it would be surprising if the space community was doing nothing about it, because debris impacts can affect the operation of unmanned spacecraft and threaten the occupants of manned spacecraft.

[NASA]

Fig. 40   A collision between NASA's DART spacecraft (top) and its target in 2005 showed that rendezvous and docking techniques have still to be perfected.

As a result of the near misses just described, space debris experts have suggested that debris tracking facilities be improved. In fact, the U.S. Air Force has already taken the initial steps by proposing to deploy a space-based space surveillance system, in the form of a constellation of satellites, to locate and track objects in low, medium, and geostationary orbits. Although the proposal is in its early stages, it is hoped that the constellation might be operational by 2012.[28]

This will do nothing to help matters in the near term, however, and the only option for most spacecraft operators is to rely on the existing network. NASA, for example, makes use of the debris tracking network to protect its Space Shuttle fleet and has incorporated specific guidance into its prelaunch flight rules. If a debris object is predicted to violate a $5 \times 25 \times 5$ km box around the vehicle during the first two hours of a mission, a launch hold is called. (The three dimensions are defined as radial up, circular downtrack, and out of plane.) As of 1997, two launches had been delayed specifically to allow larger pieces of debris to pass over the launch site.[22]

Whenever a Shuttle is *in* orbit, U.S. authorities examine the trajectories of orbital debris to identify possible close encounters, where "close" is typically defined as within 5 km (the orbital box being $2 \times 5 \times 2$ km). The consequences of impact are taken so seriously that if an object is expected to come within 2 km of the Shuttle it will normally maneuver away from it, even though the chances of a collision in this situation are less than 1 in 100,000. Records show that from STS-26 to STS-81 at least seven trackable objects entered the box, leading to avoidance maneuvers in three cases.[22] On 8 December 1992, for example, Discovery was maneuvered to avoid a 10-cm piece of debris, which could have led to the destruction of the orbiter had an impact occurred.[18]

Although technical information from the newest entrant to the field of manned spaceflight, China, is relatively thin on the ground, the potential danger to its first manned vehicle, Shenzhou 5, was considered. Although some degree of mistranslation is likely, an article in Aerospace China magazine—the mouthpiece of state-owned China Aerospace Science and Technology Corporation (CASC)—suggests that the spacecraft was fitted with "an alarm system to avoid collisions in space"[29]; according to the chief scientist at the Centre for Space Science and Applied Research, Professor Du Heng, it was designed to keep the spacecraft away from debris by "automatically changing its propulsion and speed."

Whatever the accuracy of such statements, China takes the subject of orbital debris seriously and has allocated significant funds for debris research; no doubt its plans to launch a space station will have galvanized attention on the destructive power of space debris.

Because the probability of impact is related to the surface area of a structure, the International Space Station is a special case by virtue of its size. Its overall dimensions, at completion, will be approximately 110×75 m (about the size of a soccer pitch), and each of its 16 primary solar-array panels will measure about 4.5×33 m (a total area of some 2400 m^2). Because the station and its arrays present a much larger area than any other spacecraft, the chance of being hit and damaged during its lifetime is significantly greater.

In recognition of the inevitability of debris impact, the ISS has two options. It can maneuver to avoid it—and in 2001 alone, it did so three times—or it can sit there and take it. If the object is greater than about 5 cm across, the station is maneuvered; if the object is less than a centimeter across, the ISS modules rely on their integral sacrificial shields to intercept and pulverize the debris, distributing the energy over a much wider area to avoid puncturing the module walls (see Fig. 41); but if the object is between about 1 and 5 cm across, there is nothing that can be done under the limitations of current technology.

Sacrificial shields are also known as bumper shields or Whipple shields (after Fred Whipple, the American cometary astronomer who made initial calculations on the presumed hazards of space debris to spacecraft as early as 1946). The efficacy of Whipple's meteor bumper shield technique was proved in 1986 by ESA's comet interceptor, Giotto (Fig. 12), which survived a "shot-blasting" passage through the tail of Halley's Comet. Although not by design, LDEF provided similar proof in that, even beneath blankets

a)

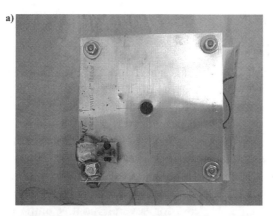

b)

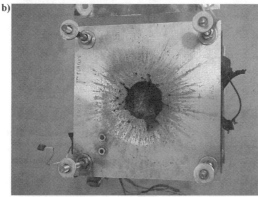

c)

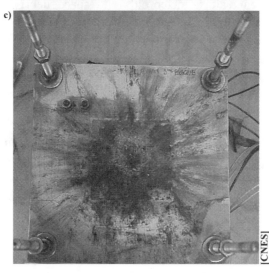

[CNES]

Fig. 41 Sacrificial shield tested by the French space agency: a) projectile entry hole in outer skin; b) center skin shows energy dissipation and wider penetration; c) final skin is damaged but not penetrated. See also Figure 12.

with 500 holes in them, it was hard to find damage to equipment: as investigators said, it all looked "brand new."

According to NASA, the ISS is the most heavily shielded spacecraft ever flown. Its habitable compartments and high-pressure tanks should normally be able to withstand the impact of debris as large as 1 cm in diameter, while the station's life-support system is designed to sustain a survivable atmosphere for the time it takes to evacuate a module that has suffered this type of impact. However, because laboratory experiments have shown that a particle just 1 cm in diameter could penetrate a 5 cm thickness of solid aluminum, the word "withstand" in this context should probably be read as "avoid being totally destroyed by." Such an incident would probably make the module unusable, at least for several months until repairs could be effected. (The disruption caused by a tiny leak in 2004 proves the point.) In fact, risk assessments conducted for the Columbus module, ESA's main contribution to the ISS, predict a 20% chance over 10 years that a particle will penetrate the shield.[18]

Unfortunately, the first two EVAs of the ISS construction sequence contributed to the debris population, when a thermal blanket, a foot restraint socket, and an access panel were lost in space. Considering that some 160 station EVAs were planned during the construction period, the potential for the creation of further debris, not to mention the exposure of astronauts to existing debris, cannot be ignored.

Although spacesuits have multiple layers, including one designed to provide protection against meteoroids and space debris, they can only protect the occupant to a point. So far, this point has not been reached, but as extravehicular activity increases so do the chances of damage to spacesuits. No one involved in manned spaceflight is under any misconceptions: space is a dangerous place and unlikely to become significantly less hazardous anytime soon.

This has serious repercussions for the future use of space stations in LEO. One of the most common, forward-looking applications for space stations involves their use as repair or construction bases, in support of manned missions to the planets for example. Unfortunately, the growing debris population means that building a large spacecraft at a space station would be hazardous, and the spacecraft itself might even be damaged before completion.

REMOVAL MECHANISM

So, is there anything that can be done to reduce the existing debris population?

At least for the lower orbits a natural clearing mechanism exists in the guise of atmospheric drag, which causes objects to reenter the atmosphere and burn up (Fig. 42). According to figures derived from debris assessment software,[30] a typical spacecraft in a circular orbit at an altitude of 250 km will reenter the Earth's atmosphere within about two months. From 600 km, an object would reenter within about 15 years, but for altitudes above 850 km, where a large proportion of satellites orbit, the decay time is typically measured in centuries. By extension, of course, a satellite in GEO will never reenter under normal circumstances.

The actual behavior of spacecraft, and smaller pieces of space debris, is complicated by their size, shape, and mass: for example, because it has a much larger surface area, and therefore suffers greater drag, a flimsy piece of insulation foil will reenter much faster than a ball bearing. Although the less dense pieces of debris are more likely to burn up completely when they do reenter, the single most significant factor in determining an object's survival is its melting temperature (Johnson, N., personal communication, 25 March 2004).

Fig. 42 Atmospheric drag causes satellites in LEO to reenter: this perspective of ESA's Eureca satellite over Cape Canaveral, Florida (from STS-46 in August 1992) gives an impression of how close to Earth some satellites are. Eureca's orbital altitude was about 500 km.

The predictability of atmospheric drag is complicated, moreover, by surges in solar activity, which expand the atmosphere and increase the drag on low-orbiting spacecraft. The variation in atmospheric density at a given height makes it more difficult to track and predict the motion of orbital debris, which has knock-on effects (possibly literally!) for spacecraft.

Although a solar-induced expansion of the upper atmosphere and resultant increase in drag is a problem for spacecraft and space-station operators, it should be good for reducing the debris population. However, it was only during the solar maximum periods of 1978–1981 and 1989–1992 that the debris catalog saw a decline in its growth rate; unfortunately, the most recent maximum in the Sun's 11-year cycle failed to produce a similar result.

Moreover, research conducted since the early 1990s has led to the suggestion that increasing carbon-dioxide (CO_2) emissions, from power stations and other terrestrial sources, are increasing the orbital lifetimes of defunct satellites and debris in the lower-altitude orbits. Observations and modeling support the theory that CO_2 emissions have led to a cooling of the thermosphere (at altitudes between 85 and 600 km) and a consequent reduction in density, which reduces the frictional effect on orbiting objects. The hypothesis is that thermospheric cooling could continue for the next 100 years, despite CO_2 emission controls, by which time the effect would be comparable with that of solar variability.[31,32]

The ultimate effect of atmospheric drag has been illustrated by a number of unplanned spacecraft reentries, including:

1) the nuclear reactor-powered Cosmos 954 satellite, which spread debris across northwest Canada in 1978;
2) the Skylab space station, which reentered over Australia in 1979;
3) the Salyut 7 space station (with docked Cosmos 1686 supply vehicle), parts of which fell on largely unpopulated areas of Argentina in 1991; and
4) the Mars 96 probe, which failed to attain an initial Earth orbit and made an uncontrolled reentry.

Launch vehicles too can produce unplanned or uncontrolled reentries, as shown by the debris recovered from a Delta second stage, which, having launched the U.S. MSX satellite, deposited fragments in Texas and Oklahoma on 22 January 1997. One particularly telling photograph shows a battered propellant tank in a field, with a not-too-distant homestead in the background. The stainless-steel tank, with a mass of a few hundred kilograms, had survived reentry virtually intact and ruptured only on hitting the ground. There have been other examples, as shown in Fig. 43.

It seems likely that objects such as this re-enter more often than we would like to think, remaining undiscovered in dense forest areas or on barren plains around the globe, with still more on the ocean floors. Standard practices

Fig. 43 Debris that reached the Earth: on 21 January 2001, the 70 kg remains of a Delta II third stage, or PAM-D, reentered the atmosphere over the Middle East and landed about 240 km from Riyadh.

devised by the U.S. government state that, if a space structure is to be disposed of by reentry into the Earth's atmosphere, the risk of human casualty should be less than 1 in 10,000. In fact, according to debris statistics, some 17,000 uncontrolled reentries of cataloged objects have occurred since the beginning of the Space Age[33]—an average of one per day—as far as we know without serious injury or significant property damage.

Despite this, these events do little to reduce the overall debris population because the majority of spacecraft reside in orbits largely unaffected by atmospheric friction. This is where a portfolio of debris mitigation measures comes into play.

MITIGATION MEASURES

Self-evidently, the best way to reduce the probability of collisions with space debris is to prevent the accumulation of debris in the first place. Although it is already too late to avoid the initial accumulation, measures are in hand to limit the growth of the debris population.

For example, it is possible to actively deorbit LEO satellites at the ends of their lives, either sending them on controlled reentries, or maneuvering them into higher or lower disposal orbits (otherwise known as reorbiting). However, this requires planning to ensure that it is done before the satellites become uncontrollable, which means allocating a certain amount of stationkeeping propellant for deorbiting maneuvers. It is currently far more common to allow the satellite orbits to decay naturally.

The French space agency CNES has a sensible attitude towards orbital debris, probably as a result of its experience with near misses already described. When control of its SPOT-3 satellite was lost in 1996, it was fated to spend the next 200 years or so gradually falling towards Earth. Not wishing to repeat the experience with SPOT-1, which was retired in mid-2002 following the launch of SPOT-5, CNES initiated plans to deorbit the satellite from its 830-km orbit in 2003, using its remaining stationkeeping fuel.[34] It is estimated that atmospheric friction will still take some 15 years to drag the satellite in, but at least this is an improvement on 200 years.

Two of the largest spacecraft to be deorbited in planned reentries were the Compton Gamma Ray Observatory, a 17-tonne satellite deorbited in June 2000, and the 135-tonne Mir space station, which reentered in March 2001.[35] Though the latter was a highly publicized political event, brought on by the Russian government's perceived inability to maintain Mir while funding its contribution to the ISS, it shows how even large space objects can be brought down relatively safely.

A problem arose with plans to deorbit the Hubble Space Telescope, however. Because it has no propulsion system of its own, it was originally intended to return it to Earth in the payload bay of a Space Shuttle. However, following the Columbia accident, NASA decided, at least initially, to pursue the development of an autonomous "space tug" that would attach itself to the HST and deorbit the telescope.[36]

Although an estimated 75% of the mass launched into orbit since the beginning of the Space Age has reentered the Earth's atmosphere, space debris experts believe that many more satellites need to be deorbited to reduce the debris population,[33] and, as the in-orbit construction of the ISS continues, the subject of debris mitigation is likely to rise higher on the agenda.

The overcrowding of *geostationary* spacecraft was highlighted by concerned individuals more than 20 years ago, and an allocation was also made in their propellant budgets for deorbiting. However, it is not yet possible to return these spacecraft to LEO because their orbit—at an altitude of 36,000 km—is too high for this to be practical. This has been solved temporarily by boosting spent satellites into what is colloquially termed a "graveyard orbit," some distance above or below GEO. The first organization to do this was Intelsat, which boosted three of its Intelsat III satellites to altitudes between 400 and 3700 km above GEO.[37]

However, according to ESA records, only one-third of geostationary satellites retired between 1998 and 2003 were moved to graveyard orbits,[34] which leaves a lot of potential debris in GEO. No doubt it would be convenient to lay the blame on "rogue operators," but even NASA has abandoned spacecraft in GEO: in 2004, it dumped its Advanced Communications Technology Satellite (ACTS) at 105.2°W because it reportedly had insufficient propellant reserves to boost it to the graveyard.[38]

In any event, although the measure leaves room in the limited geostationary orbit for the satellites' replacements, it is little more than a stopgap measure because spacecraft in these graveyard orbits are uncontrolled and themselves constitute a potential source of future space debris.

A form of graveyard or "disposal" orbit is also used for the GPS constellation, which operates in an 18,000-km-high medium Earth orbit (MEO). Unfortunately, research conducted in 2002 showed that, under the influence of gravitational and solar radiation forces, the nominally circular orbits will evolve into ellipses that will eventually intersect other orbits. For example, it is predicted that within 20–40 years the higher-altitude disposal orbits will become sufficiently eccentric to allow uncontrolled satellites to drift across the orbital paths of operational GPS spacecraft; and after 160–180 years even satellites in LEO and GEO could be threatened.[39]

Although the chances of collision, in both the near-term and long-term scenarios, are small, the potential is enough to cause concern because as many as 20 GPS satellites were expected to be retired by 2010. As a result, as of March 2002 when 14 satellites had been retired, five had been boosted to an increased separation altitude of 500 km.

Moreover, it is not only the satellites that are of concern. So far, GPS satellites have been delivered from LEO to MEO by rocket motors that remain attached to the satellite. When GPS spacecraft begin launching on the Delta IV and Atlas V Evolved Expendable Launch Vehicles (EELVs), they will be delivered to MEO by the vehicles' own upper stages from which they will be separated. This will effectively deposit unused and uncontrolled rocket stages in the vicinity of MEO, and, although they will be placed in a disposal orbit, they will be prone to the same perturbation mechanisms as the retired satellites.

To avoid the unintentional upper-stage explosions that have produced so much of existing space debris, concerned manufacturers have engineered passivation measures—essentially the venting of unused fuel. In addition, the better launch vehicles are now designed with encapsulated separation mechanisms, which allow stage and fairing separation without polluting the space environment.

However, the most effective debris mitigation techniques incur a mass penalty, in hardware or in fuel, which reduces the performance of the vehicle or the payload it can carry, and it is unrealistic to expect one manufacturer to adopt the most effective techniques if competitors are not obliged to do likewise. Despite this, U.S. manufacturers have been very responsive in support of a U.S. policy on upper-stage passivation initiated in 1981 (Johnson, N., personal communication, 4 May 2004), while Ariane stages have been passivated since the V59 mission in 1993.[40]

However, some launch providers have not made similar investments, and upper-stage fragmentation continues to be a problem. Among the more recent upper-stage breakups were that of a Long March 4 in 2000, which produced about 300 cataloged debris objects; an Indian Polar Satellite Launch Vehicle (PSLV) in 2001, creating about 325 (Johnson, N., personal communication, 25 March 2004); and a Proton Block DM in October 2004, which added more than 60 debris pieces.[41] To be fair, the LM4 incident was caused by a passivation malfunction, and the PSLV has since been redesigned, so that both vehicles are now passivated on a routine basis (Johnson, N., personal communication, 4 May 2004), but that does nothing to remove the debris. Sadly, nor does it remove the possibility of older stages breaking up, as shown by the case of the 10-year old Ariane 4 upper stage, which is believed to have created several pieces of debris in 2001.[42]

In parallel, debris mitigation measures have also been engineered for spacecraft. For example, to eliminate the generation of debris from antenna or solar-array deployment systems, non-pyrotechnic devices (which avoid the use of explosive materials) have been developed. One example is the thermal knife, a device designed to sever a cable connecting two items of hardware by heating to high temperatures. Other devices are designed to encapsulate any separation debris so that nothing is released into the space environment. In addition, because batteries on defunct spacecraft have been known to explode, it is now common to discharge them at the end of a mission. In general, the more inert a satellite can be made at the end of its life, the less trouble it is likely to be in the future.

Sometime in the indeterminate future it might be possible to remove the debris entirely, and papers proposing how this might be done are published on a regular basis. One of the most popular methods is removal by laser vaporization, which is theoretically feasible, but requires a very high-power laser. According to one author, the complete vaporization of 1 g of aluminum requires the deposition of nearly 15 kJ of energy in the material.[43] However, not all of the incident energy can be absorbed, and, depending on the laser wavelength and the temperature of the target, 80% or more can be reflected. It is therefore estimated that 100 kW of laser power would be required to remove each gram of debris, a fairly tall order in anyone's book considering that the largest satellites currently produce less than one-fifth of that power.

However, it has also been suggested that the vapor 'blown off' in the process, which creates a thrust, could be used to change the orbit of the debris and engineer reentry. But there are too many unknowns regarding the material composition of the debris to consider this a reliable procedure: for example, some materials could crack and separate rather than vaporize, while others might expel liquid that would then resolidify, adding to the debris problem.

A less energetic solution, which is perhaps nearer to our capabilities, is the automated space robot designed to dismantle and deorbit redundant hardware. One proposal features an autonomous spacecraft with remote manipulator arms and a solar-powered cutting system[44]: the spacecraft would be launched to an orbital inclination that needed clearing, store dissected pieces of defunct satellites onboard, and later deorbit them in a controlled manner. When such a system might be funded, built, and launched is, however, anyone's guess . . . and until then the problem remains.

DEBRIS EVOLUTION AND FUTURE RISKS

Despite natural clearing, deorbiting, and debris mitigation measures, the orbiting satellite population is growing and so is the risk of collisions. For example, it is estimated that there are over 100,000 pieces of debris large enough to cause the loss of a satellite, and with dozens of new satellites being launched each year this seems likely to increase.[8]

Proposals made in the 1990s for a number of satellite constellations based in LEO and MEO gave cause for concern because they were expected to increase the population almost exponentially. The original proposal for the Teledesic broadband system, for example, was based on a constellation of almost 900 satellites. Considering the absence of an agreed set of debris mitigation measures and no official body to formulate and regulate such measures, the outlook was less than optimistic.

As it turned out, however, only a small number of the proposed constellations was launched, and both Iridium and Globalstar incorporated plans to reorbit satellites—to lower and higher altitudes respectively—at the end of their useful lives. Nevertheless, the numbers of satellites involved were unprecedented: 66 for Iridium, 48 for Globalstar, and 36 for Orbcomm, not including in-orbit spares. The limited success of the respective businesses and the resultant difficulties involved in raising finance for other projects placed a temporary halt on other constellation programs, but the possibility of their resurgence remains.

A potential disaster scenario in the evolution of the debris population is the cascade effect, which predicts that when debris collisions produce large numbers of objects, those objects can undergo further collisions producing even more debris. First mooted in the 1980s, this chain reaction could lead to the closure of some of the more popular orbits within a relatively short time.[45] Although it is necessary to make a number of assumptions in this type of analysis—for example, that satellites continue to be launched to given orbits at a given rate—it has been suggested that the most likely orbital altitude for a chain reaction is between 930 and 1100 km. If launches continue as they have in the past, a critical population, which is thought to be only two to three times the current population, could be reached before the middle of the century.

Again, all predictions of the future are open to interpretation, but the suggestion that such an effect is even possible should start alarm bells ringing in the space community. It seems incredible that, by the time the Space Age is but twice its current age, some parts of space could be off limits to future explorers and developers.

Indeed, for many debris experts this scenario is too incredible: they point to increasing cooperation between spacefaring nations in mitigating the growth of space debris and the expectation that no serious degradation of the environment will occur before 2050. One wonders how long this assessment would stand in the face of a Teledesic-like constellation in LEO 10 or 15 years hence.

POLICY ASPECTS

One of the problems in controlling and limiting the proliferation of space debris in the past was that satellite and launch-vehicle manufacturers were not legally bound to employ mitigation measures and there were no internationally agreed policies. In recent years, however, a significant amount of technical study and consultation among the international space community has taken place (Fig. 44).

Much of the discussion has been facilitated by the Inter-Agency Space Debris Coordination Committee (IADC), an international governmental forum formed in 1993 to address orbital debris issues and encourage operations in Earth orbit that limit the growth of orbital debris. The IADC's members include the space agencies (or similar bodies) of Europe, France, Germany, India, Italy, Japan, the People's Republic of China, Russia, Ukraine, the United Kingdom, and the United States. Its stated primary purposes are to encourage the exchange of information on space debris among members, review cooperative activities, facilitate new opportunities for research, and identify debris mitigation options.

In addition, in 1993, the International Telecommunication Union (ITU)—a body responsible for the planning and regulation of international telecommunications services—adopted a debris mitigation recommendation, which suggested, among other things, that as little debris as possible be released into GEO, and that satellites be transferred at the end of their lives to a storage orbit that does not intersect with GEO (in other words, a graveyard orbit).

Moreover, since 1994, orbital debris has been a topic of assessment and discussion in the Scientific and Technical Subcommittee (STSC) of the United Nations' Committee on the Peaceful Uses of Outer Space (UNCOPUOS). In 1999, the STSC adopted a comprehensive report on orbital debris,[46] which concluded that "the known and assessed population of debris is growing, and the probabilities of potentially damaging collisions

Fig. 44 A great deal of technical information on space debris has been published ... but more work remains to be done on the policy aspects.

will consequently increase. [Therefore] the implementation of some debris mitigation measures today is a prudent step towards preserving space for future generations."

As a result, the IADC developed a set of debris mitigation guidelines, which were approved by the world's leading space agencies and submitted to the United Nations in late 2002. Following a formal presentation to the STSC in February 2003, it was hoped that the guidelines would be endorsed by the United Nations, but negotiations reached an impasse at the 2004 meeting. However, in February 2005 the first draft U.N. guidelines were completed and seem likely to be adopted by COPUOS in 2007 (Johnson, N., personal communication, 26 June 2005).

On a national as opposed to international level, the United States has been particularly active in the discussion of debris. This is nothing less than one would expect, considering the importance of the United States as a leading space nation. As in many other walks of life, where America goes, others follow.

Since 1989, when the U.S. government released its first report on space debris, the "Report on Orbital Debris for the National Security Council," the official policy of the United States has been to minimize the creation of new orbital debris. Its "1995 Interagency Report," drafted under the direction of the White House Office of Science and Technology Policy, went further and recommended that NASA and the U.S. Department of Defense (DoD) jointly develop draft design guidelines and consult with the private sector in applying those guidelines to future satellite systems. This recognition of the need to include the space industry in any discussions was crucial to the adoption of any future policy, because it would be industry that produced the majority of future satellites.

Both the 1995 Interagency Report and the 1996 National Space Policy recognized that there were important *international* aspects of the debris mitigation policy. The former noted the need for a coordinated U.S.-international strategy to encourage other nations to adopt debris policies and practices, while the latter indicated that it is in the interest of the U.S. government to ensure that space debris minimization practices are applied by other spacefaring nations. Obviously, debris produced by other nations' satellites could damage U.S. spacecraft as well as their own.

The National Space Policy also stated, unsurprisingly, that the U.S. government would take a leadership role in international fora. As a result, in January 1998, draft U.S. government guidelines were presented at a government workshop for industry. The practices, which have since been adopted as Government Standard Practices, include controlling debris released during normal operations, minimizing debris generated by accidental explosions, and postmission disposal of space structures.

If a spacecraft cannot be safely reentered, the guidelines suggest the use of three main storage orbits:

1) between LEO and MEO (perigee above 2000 km and apogee below 19,700 km),
2) between MEO and GEO (20,700–35,300 km), and
3) above GEO (perigee above 36,100 km, or approximately 300 km above GEO).

A fourth suggestion is to remove the spacecraft from Earth orbit by maneuvering it into a heliocentric orbit (an orbit around the Sun).

However, although the Government Standard Practices have been adopted for U.S. government missions, government agencies retain the right to deviate from the specific practices "if necessary to address considerations of cost or mission effectiveness."[30] Moreover, the practices are not directly applicable to nongovernment missions, so they are a long way from being hard and fast rules, let alone laws.

Nevertheless, common sense has gone some way to alleviate the situation. For example, in the United States, licensing authority for *nongovernment* space activities rests with three agencies: the Federal Aviation Administration (FAA), the National Oceanic and Atmospheric Administration (NOAA), and the Federal Communications Commission (FCC). The FAA licences commercial launches and regulates U.S. launch activities, NOAA licenses commercial remote sensing systems under the Land Remote Sensing Policy Act of 1992, and the FCC licenses telecommunications services within the United States and between the United States and foreign carriers.

Both the FAA and NOAA have issued regulations concerning the mitigation of orbital debris that require nongovernmental space missions to adopt practices consistent with the Government Standard Practices. Specifically, the FAA requires applicants for commercial launch licenses to demonstrate that no launch-vehicle stages, components, or payloads that reach Earth orbit will suffer "unintended physical contact." In addition, it requires the passivation of propellant tanks, batteries, and other energy sources that might cause fragmentation of a vehicle or its components.

For its part, the FCC did not feel it necessary to make debris mitigation plans until it began to license the LEO constellations in the 1990s, when it was felt by many observers, within and beyond the FCC, that multiple-satellite constellations presented a significant increase in the risk of collision and fragmentation debris. The FCC has since proposed to extend its recommendations to satellites in MEO and GEO.

Specifically, in May 2002, the FCC issued a Notice of Proposed Rule Making (NPRM) on the mitigation of orbital debris.[30] Intended partly as a consultation document, the NPRM was designed to help the FCC decide how debris mitigation issues should be incorporated into its rules and licensing procedures.

It stated that:

> [although] the immediate risk presented by orbital debris is minimal, prudent measures adopted now are important to ensure continued affordable access to space, the continued provision of reliable space based communications services, and the continued safety of persons and property on the surface of the Earth. Orbital debris mitigation measures are therefore an important part of [satellite operations licensed by the FCC].[30]

As a result, the NPRM proposed, as part of the licensing process, the "disclosure of orbital debris mitigation plans for all types of satellite systems licensed by the FCC."

QUESTIONS

Although the space community is making headway towards the solution to the orbital debris problem, many questions remain—not least regarding the optimum height for the graveyard orbit (see Box: The Altitude of the Graveyard Orbit).

The fact that many issues are far from cut and dried was indicated by the questioning nature of the FCC's NPRM. For example, it questioned whether the quality of LEO constellation spacecraft could be ensured, whether satellite operators should be obliged to obtain insurance to cover debris damage, and whether their debris mitigation measures should be assessed before a license was granted. In principle, any measure that reduces the potential for debris seems like a good idea, but the cost of the measures will always be an issue, even if recommendations are, one day, replaced by laws.

The FCC also asked whether it should be mandatory for operators to report their end-of-life propellant margins and whether spacecraft maneuvers should be formally coordinated, similar to the way aircraft maneuvers are coordinated by air-traffic-control authorities. Though operators might not welcome the extra bureaucracy of margin reports, and current U.S. export controls on satellite information would make it difficult for U.S. operators to share data with foreign entities, this might provide a degree of confidence that satellites were capable of being removed from GEO.

Maneuvers are coordinated on an *ad hoc* basis at present because relatively few spacecraft conduct regular maneuvers beyond stationkeeping: in general, unmanned spacecraft maintain their orbital parameters until they reach the end of their lives, or are relocated to another geostationary orbital position. This suggests that the current system can be retained in the near term; however, once orbital transfer vehicles (colloquially termed space tugs) are developed and deployed, the situation might have to be reassessed, because their raison d'être is to maneuver.

Finally, the NPRM asked whether the FCC rules should cover foreign satellites and launch vehicles. This provides a timely reminder that space is an international regime, not one controlled by U.S. agencies however capable and well meaning, and argues for an international policy. As one law professor put it, "Space is an area that, with the possible exception of geostationary orbit, is simply not amenable to the traditional kinds of jurisdictional line-drawing. To that extent, it physically must be shared by all nations."[47]

The policies proposed by the IADC are perhaps the closest to this ideal, but the fact that individual national space and space-related agencies, including at least four in the United States alone, are developing their own debris mitigation policies makes agreement difficult. Some debris experts believe that a United Nations resolution would help, as it has with the regulation of nuclear power in orbit (the "U.N. Principles on the Use of Nuclear Power Sources in Outer Space," as adopted in 1992), but no one thinks that international regulations are easy either.

Even when nations agree in principle, there is still the question of liability in specific situations (Fig. 45). For example, if a debris cloud from one spacecraft causes damage to another, whose responsibility is it? Who pays for the damage? Is it up to an owner to insure against damage from another spacecraft or, to borrow a phrase from terrestrial environmentalism, should the polluter pay? Of course, the situation for space is quite different from that for Earth: if we stop polluting a terrestrial stream, it can recover in a few years; if we stop producing space debris, very little regeneration takes place.[48]

Is legislation the answer? Should protection of the space environment be an integral part of the body of space law? Or should the space community rely on a form of "gentlemen's agreement" in much the same way as radio-frequency interference is minimized today under the auspices of the ITU?

These and other questions will have to be answered sooner or later ... and sooner would be better. A statement by FCC commissioner Michael J. Copps attached to the NPRM summed it up nicely:

> It is important ... that we sometimes look over the horizon and try to head off problems before they occur, rather than waiting for the problems to find us unprepared. An ounce of prevention is worth a pound of cure, and if we come up with the right orbital debris mitigation rules now, we can head off a potentially very costly problem with far less costly precautions.[30]

[Willis Inspace]

Fig. 45   A question of liability: if one spacecraft damages another, is the owner liable?

In the early days of the Space Age, there was a somewhat cavalier attitude towards the use (and abuse) of orbits, but it is now widely understood and agreed within the space community that the orbital environment is worthy of protection. This understanding is based on observations which show that debris in some of the most important orbits is increasing. The agreement has been reached for purely pragmatic reasons, in that failure to do so would eventually lead to these orbits being closed to further development and exploitation.

There might still be more questions than answers regarding debris in Earth orbit, but at least the debate has begun. But what about the other planets of the solar system and their orbits? This is the topic of the next chapter.

THE ALTITUDE OF THE GRAVEYARD ORBIT

At the end of their useful lives, most geostationary satellites are boosted from their operational orbits, at an average altitude of 35,786 km, to a graveyard orbit. Originally, both ITU and U.S. Government Standard Practices suggested that the graveyard should be 300 km above GEO. In contrast, the IADC recommended a range that varies from 235 to 435 km, dependent on the satellite, implying that the lower figure provides adequate separation.

The magnitude of the figure is far from academic because all maneuvers require the expenditure of propellant that could otherwise be used for stationkeeping: it is estimated that every additional 25 km equates to a week of revenue-earning lifetime.[49] This means that some three months-worth of stationkeeping would be sacrificed to boost the satellite to a 300 km graveyard. With transponder lease rates averaging $1 million per year and most satellites carrying 20 or 30 transponders, the financial impact becomes clear.

Considering the difficulties of the satcoms business, which include high entry costs, thin profit margins, sceptical investors, and growing terrestrial competition, the last thing operators want to do is deactivate a perfectly good satellite before they need to.[49] Nevertheless, many satellites are being retired successfully to the graveyard. In 1997–98, for example, 38 satellites were retired, of which 13 exceeded the minimum IADC altitude of 235 km and nine exceeded the US/ITU altitude of 300 km. Unfortunately, the other 12 (11 of which were either Chinese or Russian in origin) were abandoned in GEO; the American Telstar 401 suffered an unexpected total failure and could not be removed.[49]

However, later figures show that the recommendation for removal has not been followed by some operators. According to one author,[50] it has been complied with by "only two out of 14 satellites [approaching] the ends of their active lives in 2001" and from 1996 to 2000, "a total of 50 satellites were left in or near the nominal geostationary orbit, posing a risk of collision with one or other of the active satellites." Likewise, IADC figures show that, in 2002, only five of 13 and, in 2003, only six of 15 retired satellites were removed to graveyard orbits.[51] It seems that although many operators are behaving responsibly with regard to satellite disposal, some are not.

Not content with its previous recommendation, the IADC later developed a more detailed formula for deriving the graveyard altitude, which takes into account a number of factors that might affect the long-term stability of the orbit. It suggested an orbit with a perigee altitude of no less than $36{,}021\,\mathrm{km} + (1000\,C_R A/m)$, where C_R is the solar radiation pressure coefficient of the spacecraft (between 0 and 2); A/m is the area-to-mass ratio of the spacecraft in square meters per kilogram; and 36,021 is $35{,}786 + 235$.

Although many observers believe that using a single number like 300 km is much simpler and as efficient, the ITU now uses the IADC formula for computing the altitude increase (Flury, W., personal communication, 25 May 2004). Meanwhile, based on IADC recommendations and following several years of discussion and consultation, the FCC published its ruling on 21 June 2004. Despite the objections of several commercial satellite operators, it ruled that all U.S.-licensed satellites launched after 18 March 2002 will have to be placed in graveyard orbits between 200 and 300 km above GEO at the end of their lives. This is a condition for granting an operating license for the United States.[51] In October 2005, the FCC extended this to require all license applicants to include detailed commitments to debris mitigation in their applications.

REFERENCES

[1]*Collins Concise English Dictionary*, 3rd ed., Harper Collins, Glasgow, Scotland, UK, 1992, p. 335.

[2]Williamson, M., *The Cambridge Dictionary of Space Technology*, Cambridge Univ. Press, Cambridge, England, UK, 2001, p. 227.

[3]Williamson, M., *The Cambridge Dictionary of Space Technology*, Cambridge Univ. Press, Cambridge, England, UK, 2001, p. 351.

[4]Flury, W., "Space Debris: a Report from the ESA Space Debris Working Group," European Space Agency, SP-1109, Paris, France, Nov. 1988.

[5]"ESA Space Debris Mitigation Handbook," Release 1.0, cover dated 7 April 1999; issued on 19 Feb. 1999; 2nd ed. dated 3 March 2003.

[6]Anz-Meader, P. D., Johnson, N., Cizek, E., and Portman, S., *History of On-Orbit Satellite Fragmentations*, 12th ed., NASA Lyndon B. Johnson Space Center Orbital Debris Program Office, Houston, TX, JSC29517, 31 July 2001.

[7]Kessler, D. J., "Earth Orbital Pollution," *Beyond Spaceship Earth: Environmental Ethics and the Solar System*, edited by E. C. Hargrove, Sierra Club Books, San Francisco, 1986, p. 54.

[8]Ailor, W., "Meteoroids, Space Debris, and Related Space Hazards," Analytical Graphics Insurance, Nov. 1998.

[9]Yasaka, T., Hanada, T., and Matsuoka, T., "Model of the Geosynchronous Debris Environment," *International Astronautical Congress*, Paper 96-IAA.6.3.08, Oct. 1996.

[10]"Military Space; Country: International," *Jane's Space Directory 1997–98*, Jane's Publishing, London, 1997, p. 146.

[11]Gatland, K. W., *Astronautics in the Sixties*, Iliffe Books, Ltd., London, 1962, pp. 60,145.

[12]Rusk, D., "Star Wars: The Nuclear/Military Uses of Space," *Beyond Spaceship Earth: Environmental Ethics and the Solar System*, edited by E. C. Hargrove, Sierra Club Books, San Francisco, 1986, pp. 317–318.

[13]Mehrholtz, D., Leushacke, L., Flury, W., Jehn, R., Klinkrad, H., and Landgraf, M., "Detecting, Tracking and Imaging Space Debris," *ESA Bulletin*, Vol. 109, Feb. 2002, pp. 128–134.

[14]David, L., "Havoc in the Heavens: Soviet-Era Satellite's Leaky Reactor's Lethal Legacy," Space.com [cited 29 March 2004].

[15]Rains, L., "Long-Delayed Russian Optical Tracking Complex Is Open," *Space News*, Vol. 13, 22 July 2002, p. 4.

[16]Singer, J., "US Air Force's SAINT Will Study Satellites Close to Earth," *Space News*, Vol. 15, 26 April 2004, p. 17.

[17]Liou, J.-C. (ed.), "Orbital Box Score," *The Orbital Debris Quarterly News*, Vol. 9, No. 2, April 2005, p. 10.

[18]Flury, W., "Space Debris: An Overview," *Earth Space Review*, Vol. 9, No. 4, 2000, pp. 40–47.

[19]Drolshagen, G., "In-Orbit Measurements of Particulates," Space Hazards Workshop, DERA, Oct. 1998.

[20]Kessler, D. J., "Earth Orbital Pollution," *Beyond Spaceship Earth: Environmental Ethics and the Solar System*, edited by E. C. Hargrove, Sierra Club Books, San Francisco, 1986, p. 57.

[21]De Selding, P., *Space News*, 24–30 March 1997, p. 3.

[22]Jensen, T., Loyd, S., Whittle, D., and Christiansen, E., "Visible Effects of Space Debris on the Shuttle Program," International Astronautical Federation, Paper 97-IAA.6.4.02, 1997.

[23]Harland, D. M., *The Mir Space Station*, Wiley-Praxis, Chichester, England, UK, 1997, pp. 208,209.

[24]Bonnal, C., "Ariane's Third Stage Passivation System," CNES dossier on 'Orbital Debris,' *CNES Magazine*, No. 4, Jan. 1999, pp. 21,22.

[25]Sweeting, M., Hashida, Y., Bean, N. P., Hodgart, M. S., and Steyn, H., "Cerise Microsatellite Recovery from First Detected Collision in Low Earth Orbit," International Astronautical Federation, Paper 97-IAA.6.4.03, Oct. 1997.

[26]David, L., "U.S.-China Space Debris Collide in Orbit," Space.com [cited 16 April 2005].

[27]Alby, F., "Collision Risk Monitoring for SPOT Satellites," Space Hazards Workshop, DERA, Oct. 1998.

[28]"USAF Planning Space Surveillance System," *Space News*, Vol. 13, 12 Aug. 2002, p. 11.

[29]Xue, F., "Second National Special Symposium on Space Debris Held in Shanghai," *Aerospace China*, Vol. 4, No. 3, Autumn 2003, p. 22.

[30]"In the matter of Mitigation of Orbital Debris," Federal Communications Commission, Notice of Proposed Rule Making, IB Docket No. 02-54, adopted 14 March 2002; released 18 March 2002.

[31]Emmert, J. T., Picone, J. M., Lean, J. L., and Knowles, S. H., "Global Change in the Thermosphere: Compelling Evidence of a Secular Decrease in Density," *Journal of Geophysical Research*, Vol. 109, A02301, 2004.

[32]Lewis, H. G., Swinerd, G. G., Ellis, C. S., and Martin, C. E., "Response of the Space Debris Environment to Greenhouse Cooling," European Space Agency, Fourth European Conference on Space Debris at the European Space Operations Centre (ESOC), April 2005.

[33]Magnuson, S., "Aerospace Corp. Redeveloping 'Black Box' for Satellites," *Space News*, 18 Nov. 2002, p. 9.

[34]De Selding, P., "CNES Begins Deorbiting Spot-1 Earth Observation Satellite," *Space News*, 24 Nov. 2003, p. 6.

[35]Klinkrad, H. et al., "ESOC Activities During the Mir De-Orbit," *Proceedings of the International Workshop 'Mir Deorbit,'* European Space Agency, Paris, France, 2002, pp. 57–66.

[36]Berger, B., "NASA Proposes $300 Million Tug to Deorbit Hubble," *Space News*, Vol. 14, 24 Nov. 2003, p. 6.

[37]Perek, L., "Safety in the Geostationary Orbit," International Astronautical Federation, Paper IAA-89-632, Oct. 1989.

[38]Berger, B., "Lack of Funding Leads to Shutdown of ACTS Satellite," *Space News*, Vol. 15, 3 May 2004, p. 3.

[39]Bates, J., "Study Recommends New Plan for Retired GPS Craft," *Space News*, Vol. 13, 25 March 2002, p. 18.

[40]Bonnal, C., and Naumann, W., "Ariane Debris Mitigation Measures—Past and Future," International Astronautical Federation, Paper IAF-96-V.6.02, Oct. 1996.

[41]Liou, J.-C. (ed.), "Accidental Collisions of Cataloged Satellites Identified," *The Orbital Debris Quarterly News*, Vol. 9, No. 2, April 2005, pp. 1,2.

[42]David, L., "Space Junk and ISS: A Threatening Problem," Space.com [cited 7 Jan. 2002].

[43]Schall, W., "Orbital Debris Removal by Laser Radiation," International Astronautical Federation, Paper IAA-90-569, Oct. 1990.

[44]Ramohalli, K., and Jackson, J., "Space Debris: An Engineering Solution with an Autonomous Space Robot," International Astronautical Federation, Paper IAA-94-IAA.6.5.695, Oct. 1994.

[45]Eichler, P., and Rex, D., "Chain Reaction of Debris Generation by Collisions in Space—A Final Threat to Spaceflight?," International Astronautical Federation, Paper IAA-89-628, Oct. 1989.

[46]"Technical Report on Space Debris," United Nations, A/AC.105/720, New York, 1999.

[47]Wood, D. P., "Who Should Regulate the Space Environment?" *Preservation of Near-Earth Space for Future Generations*, edited by J. A. Simpson, Cambridge Univ. Press, Cambridge, England, UK, 1994, pp. 189–197.

[48]Williamson, R. A., "Orbital Debris—A Continuing Concern," *Earth Space Review*, Vol. 9, No. 4, 2000, pp. 30–32.

[49]Dalbello, R., "Cash Cows and Space Debris," *Via Satellite*, Vol. 17, June 2002, p. 38.

[50]Perek, L., "Planetary Protection: Lessons Learned," IAF World Space Congress, COSPAR02-A-00371, Oct. 2002.

[51]De Selding, P., "FCC Imposes Strict New Rules for Disposing of Commercial Satellites," *Space News*, Vol. 15, 28 June 2004, pp. 1,3.

Chapter 4

PLANETARY BODIES

Space debris in orbit around the Earth is now recognized as a problem for current and future users of the space environment, but relatively little consideration has been given to the volume of space beyond Earth orbit.

We have spent some 50 years exploring the solar system, beginning with near-Earth space and the Moon, and then extending our reach to the rest of the planets. Although much has been achieved by remote sensing, using the Hubble Space Telescope and other orbiting observatories, by far the most detailed information has been gained by sending spacecraft to the planetary bodies in question and using what, in medical terminology, would be described as invasive techniques.

In the early years of the Space Age, the exploration of the solar system was considered a triumph of mankind over nature described in terms of "the conquest of space." However, it also resulted in a considerable amount of space-craft debris and a number of spent rocket stages being crashed onto the surface of the Moon or abandoned in lunar orbit.[1] To a large extent, it remains "the forgotten space debris."

This chapter looks at the detrimental effects of space exploration on the planetary bodies and considers the consequences of future development.

LUNAR IMPACT HISTORY

The first stage of lunar exploration involved the crude technique of launching hardware to impact the Moon, an occupation in which success was measured by how close the spacecraft got to the target. It was only later that the more advanced techniques for orbiting and landing were perfected.

The first spacecraft to hit the Moon, on 13 September 1959, was the Soviet Union's Luna 2 (Fig. 46). It carried a 26-kg sphere, which, according to one contemporary account, disintegrated on impact, scattering tiny medallions imprinted with Russia's hammer-and-sickle emblem over the surface of the Moon.[2] In fact, no one really knows what happened to the spacecraft, the sphere, and the rocket upper stage that carried them there. And we shall not know conclusively until the impact area is located and imaged at high resolution, or a landing party investigates the crater(s) produced by

Fig. 46   Luna 2, the first spacecraft to hit the Moon.

the impact(s). Depending on the velocity of the hardware and the constitution of the surface, the force of the impact could have buried the sphere and its medallions under several meters of crust, fused them together in an unrecognizable mass, or spread them in a fountain of material across the surface.

In the same way that we have limited information on the debris created in orbit around the Earth, we have an incomplete knowledge of the debris created by spacecraft impacts on the lunar surface. And if our knowledge of the Moon's near side is incomplete, what about the far side? Because the Moon is in captured rotation about the Earth, it presents the same face towards us at all times. So, apart from the 9% of the far side revealed by a slight wobble known as libration, the only way to view it is from a spacecraft.

The first spacecraft to crash on the far side of the Moon, in April 1962, was America's Ranger 4. Unfortunately, it was inactive and failed to provide data because the master clock in its central computer had stopped.[3]

In the three years that followed, Ranger 6, 7, 8, and 9, and Luna 5, 7, and 8 were all crashed onto the lunar surface (see Table 3). Later, in an attempt to upstage Apollo 11, the Soviets sent Luna 15 to the Moon: it crashed in the Mare Crisium (the aptly named Sea of Crises) on 21 July 1969, the day after Apollo 11 landed safely in the adjacent Mare Tranquillitatis (Sea of Tranquillity).

The intention of the Ranger missions, as far as NASA and the U.S. space science community were concerned, was to photograph the Moon at close range in preparation for the manned Apollo program (Fig. 47). Any suggestion that littering the surface with the debris of crashed spacecraft was tantamount to pollution of the space environment was not taken seriously by most people; it was simply a necessary step in the exploration of an alien world.

Moreover, if Apollo astronauts were one day to land safely on the Moon, NASA had to prove that a spacecraft could be soft landed on the surface— and this was the aim of the Surveyor program. When Surveyor 1 made the first successful soft landing in June 1966, it was hailed as a success for NASA; incidentally, it was also a good result for the Moon because the amount of debris was limited. Debris was not, however, *eliminated* because a successful soft landing of a Surveyor spacecraft required the use, and subsequent jettisoning of, a separate propulsion unit, which later impacted the Moon.

The situation was similar for the Soviet Luna spacecraft, from which additional debris objects were separated, and the very first lunar impact— of Luna 2—was accompanied by the separate impact of its Vostok launch-vehicle upper stage.[4] Other missions prior to Luna 9's successful landing failed for a variety of reasons, including the apparent inability to fire retro-rockets at the correct time: on Luna 7 they fired too early, and the probe fell uncontrolled to the Moon, whereas on Luna 8 they fired too late and it smashed itself into the surface.[5]

By the time of Surveyor 1's landing then, the lunar surface had received not only the debris of intentionally impacted spacecraft, but also that of a number of failed soft-landing attempts.

Although these were genuine mistakes, the policy of intentionally crashing spacecraft on the Moon was still in force in the late 1960s, when NASA ended the successful missions of its Lunar Orbiter spacecraft by command-ing them to crash (or, in the case of Lunar Orbiter 4, allowing it to deorbit).[4] This was done to "prevent interference" with the subsequent Apollo mis-sions. Although the policy seemed sensible at the time, considering that astronauts' lives were potentially at stake, the Lunar Orbiter series produced a substantial amount of lunar spacecraft debris (Fig. 48).

This paled into insignificance, however, once manned flights to the Moon began. As part of the mission plan, it was decided to deorbit the Apollo lunar-module (LM) ascent stage once the astronauts had returned to the command and service module (CSM) in lunar orbit. As a result, the ascent stages of Apollo 11, 12, 14, 15, 16, and 17—each weighing over a tonne and therefore

Table 3   Lunar and planetary impact catalog[a]

Object[b]	Launch date[c]	Latitude	Longitude	Impact nature/date	Other debris/notes[d]
			Lunar impacts		
Luna 2	12/9/1959	30.0 N	0.0 E	Crash 13/9/1959	Luna 2 upper stage
Ranger 4	23/4/1962	15.5 S	130.5 W	Crash 26/4/1962	
Ranger 6	30/1/1964	9.2 N	21.5 E	Crash 2/2/1964	—
Ranger 7	28/7/1964	10.7 S	20.7 W	Crash 31/7/1964	—
Ranger 8	17/2/1965	2.7 N	24.8 E	Crash 20/2/1965	—
Ranger 9	21/3/1965	12.9 S	2.4 W	Crash 24/3/1965	—
Luna 5	9/5/1965	31.0 S	8.0 W	Crash 12/5/1965	Propulsion unit + 2 DO[e]
Luna 7	4/10/1965	9.0 N	49.0 W	Crash	Propulsion unit + 2 DO
Luna 8	3/12/1965	9.1 N	63.3 W	Crash	Propulsion unit + 2 DO
Luna 9	31/1/1966	7.1 N	64.4 W	Hard landing 3/2/1966	Propulsion unit + 2 DO
Surveyor 1	30/5/1966	2.5 S	43.2 W	Soft landing 2/6/1966	Propulsion unit
Lunar Orbiter 1	10/8/1966	6.7 N	162.0 E	Crash (deorbited 29/10/1966)	Propulsion unit + 1 DO?
Surveyor 2	20/9/1966	5.5 N	12.0 W	Crash	
Lunar Orbiter 2	7/11/1966	4.0 S	98.0 E	Crash (deorbited 11/10/1967)	Propulsion unit + 2 DO
Luna 13	21/12/1966	18.9 N	60.0 W	Hard landing 24/12/1966	
Lunar Orbiter 3	25/2/1967	14.6 N	91.7 W	Crash (deorbited 1967)	Propulsion unit + 1 DO
Surveyor 3	17/4/1967	3.2 N	23.4 W	Soft landing 20/4/1967	
Lunar Orbiter 4	4/5/1967	—	—	Crash (decayed 1967)	
Surveyor 4	14/7/1967	0.4 N	1.3 W	Crash?	Propulsion unit + 1 DO?
Lunar Orbiter 5	1/8/1967	2.8 S	83.0 W	Crash (deorbited 31/1/1968)	
Surveyor 5	8/9/1967	1.5 N	23.2 E	Soft landing	Propulsion unit + 1 DO
Surveyor 6	7/11/1967	0.5 N	1.4 W	Soft landing 10/11/1967	Propulsion unit + 1 DO
Surveyor 7	17/1/1968	41.0 S	11.4 W	Soft landing	Propulsion unit + 1 DO
Apollo 10 LM DS[f]	18/5/1969	—	—	Crash (1969?)	
Apollo 11 LM DS	16/7/1969	0.7 N	23.5 E	Soft landing 20/7/1969	Experiments and flag

Apollo 11 LM AS[g]	16/7/1969	—	—	Crash (decayed 1969)	—
Luna 15	13/7/1969	17.0 N	60.0 E	Crash (deorbited 21/7/1969)	—
Apollo 12 LM DS	14/11/1969	3.0 S	23.4 W	Soft landing 19/11/1969	Experiments and flag
Apollo 12 LM AS	14/11/1969	5.5 S	23.4 W	Crash (deorbited 20/11/1969)	—
Apollo 13 S-IVB[h]	11/4/1970	2.4 S	27.9 W	Crash	—
Luna 16 DS	12/9/1970	0.7 S	56.3 E	Soft landing 20/9/1970	(First sample return mission)
Luna 17 DS	10/11/1970	38.3 N	35.0 W	Soft landing 17/11/1970	Carried Lunokhod 1
Lunokhod 1	10/11/1970	38.3 N	35.0 W	Deployed from Luna 17	—
Apollo 14 S-IVB	31/1/1971	8.0 S	26.6 W	Crash	—
Apollo 14 LM DS	31/1/1971	3.6 S	17.5 W	Soft landing 31/1/1971	Experiments, flag, cart, 2 golf balls (>100-m range)
Apollo 14 LM AS	31/1/1971	3.5 S	19.3 W	Crash (deorbited 7/2/1971)	—
Apollo 15 S-IVB	26/7/1971	1.0 S	11.9 W	Crash	—
Apollo 15 LM DS	26/7/1971	26.1 N	3.6 E	Soft landing 30/7/1971	Experiments and flag (carried LRV[i])
Apollo 15 LRV[i]	26/7/1971	26.1 N	3.6 E	Deployed from LM	—
Apollo 15 LM AS	26/7/1971	26.4 N	0.3 E	Crash (deorbited 3/8/1971)	—
Apollo 15 Subsat	7/1972	—	—	Crash (decayed 1973)	—
Luna 18	2/9/1971	3.6 N	56.5 E	Crash (deorbited 1971)	—
Luna 20 DS	14/2/1972	3.5 N	56.6 E	Soft landing	—
Apollo 16 S-IVB	16/4/1972	1.8 N	23.3 W	Crash	—
Apollo 16 LM DS	16/4/1972	9.0 S	15.5 E	Soft landing 21/4/1972	Experiments and flag (carried LRV[i])
Apollo 16 LRV[i]	16/4/1972	9.0 S	15.5 E	Deployed from LM	—
Apollo 16 LM AS	16/4/1972	—	—	Crash (decayed 29/5/1972)	—
Apollo 16 Subsat	4/1972	10.2 N	111.9 E	Crash (decayed 1972)	—
Apollo 17 S-IVB	7/11/1972	4.2 S	12.3 W	Crash	—

(Continued)

Table 3 Lunar and planetary impact catalog[a] (continued)

Object[b]	Launch date[c]	Latitude	Longitude	Impact nature/date	Other debris/notes[d]
Apollo 17 LM DS	7/11/1972	20.2 N	30.8 E	Soft landing 11/12/1972	Experiments and flag (carried LRV[i])
Apollo 17 LRV[i]	7/11/1972	20.2 N	30.8 E	Deployed from LM	—
Apollo 17 LM AS	7/11/1972	20.0 N	30.7 E	Crash 15/12/1972 (deorbited 1972)	—
Luna 21 DS	8/1/1973	25.9 N	30.5 E	Soft landing 16/1/1973	Carried Lunokhod 2
Lunokhod 2	8/1/1973	25.9 N	30.5 E	Deployed from Luna 21	—
Luna 23	28/10/1974	13.0 N	62.0 E	Soft landing	—
Luna 24 DS	9/8/1976	12.8 N	62.2 E	Soft landing 18/8/1976	—
Muses A/Hiten	24/1/1990	38.0 S	5.0 E	Crash (deorbited 11/4/1993)	(released from Hiten in lunar orbit)
Hagoromo	19/3/1990	—	—	Crash assumed 4/1993	—
Lunar Prospector	6/1/1998	88.0 S	45.0 W	Crash (deorbited 31/7/1999)	—
Planetary impacts: Venus[j]					
Venera 3 Capsule	16/11/1965	—	—	Hard landing 1/3/1966	Capsule cover[k]
Venera 4 Capsule	12/6/1967	—	—	Hard landing 18/10/1967	Capsule cover[k]
Venera 5 Capsule	5/1/1969	—	—	Hard landing 16/5/1969	Capsule cover[k]
Venera 6 Capsule	10/1/1969	—	—	Hard landing 17/5/1969	Capsule cover[k]
Venera 7 Capsule	17/8/1970	—	—	Hard landing 15/12/1970	Capsule cover[k]
Venera 8 Capsule	26/3/1972	—	—	Hard landing 22/7/1972	Capsule cover[k]
Venera 9 Lander	8/6/1975	32 N	69 W	Soft landing 21/10/1975	2 halves of protective sphere
Venera 10 Lander	14/6/1975	16 N	69 W	Soft landing 25/10/1975	2 halves of protective sphere
Venera 11 Lander	9/9/1978	13 S	60 W	Soft landing 21/12/1978	2 halves of protective sphere
Venera 12 Lander	14/9/1978	7 S	66 W	Soft landing 25/12/1978	2 halves of protective sphere
Pioneer Venus 2 bus	8/8/1978	33 S	70 W	Crash	—
P'r Venus 2 Probe 1	8/8/1978	0	43 W	Hard landing	—
P'r Venus 2 Probe 2	8/8/1978	75 N	20 E	Hard landing	—
P'r Venus 2 Probe 3	8/8/1978	26 S	45 W	Hard landing	—

P'r Venus 2 Probe 4	8/8/1978	27 S	45 E	Hard landing	—
Venera 13 Lander	30/10/1981	8 S	57 W	Soft landing	2 halves of protective sphere
Venera 14 Lander	4/11/1981	13 S	50 W	Soft landing	2 halves of protective sphere
Vega 1 Lander	15/12/1985	7 N	178 E	Soft landing 11/6/1985	2 halves of protective sphere
Vega 1 Aerostat	15/12/1985	—	—	Soft landing 9/6/1985	(Balloon)
Vega 2 Lander	21/12/1985	6 S	179 W	Soft landing 15/6/1985	2 halves of protective sphere
Vega 2 Aerostat	21/12/1985	—	—	Soft landing 15/6/1985	(Balloon)
Magellan	4/5/1989	—	—	Crash 13/10/1994	Probably destroyed during entry

Planetary impacts: Mars

Mars 2 Lander	19/5/1971	45 S	58 E	Hard landing 27/11/1971	4 DO
Mars 3 Lander	28/5/1971	45 S	160 W	Hard landing 2/12/1971	4 DO
Mars 6 Lander	5/8/1973	24 S	110 E	Hard landing 12/3/1974	4 DO
Viking 1 Lander	20/8/1975	22 N	48 W	Soft landing 20/7/1976	—
Viking 2 Lander	9/9/1975	48 N	134 E	Soft landing 3/9/1976	—
Mars Pathfinder Lander	4/12/1996	19 N	34 W	Soft landing 4/7/1997	3 DO (airbags)
Mars Pathfinder Rover	4/12/1996	19 N	34 W	Soft landing 4/7/1997	Rover: Sojourner
Mars Climate Orbiter	11/12/1998	—	—	Crash? 23/9/1999	Comms. lost before entering orbit
Mars Polar Lander	3/1/1999	—	—	Crash? 3/12/1999	Carried Deep Space 2 microprobes
MER Lander-A[1]	10/6/2003	14.5 S	175 E	Soft landing 3/1/2004	Mars Exploration Rover-A (airbags)
Spirit Rover	10/6/2003	14.5 S	175 E	Soft landing 3/1/2004	Carried by MER-A (+ other DO)
MER Lander-B	7/7/2003	2 S	354 E	Soft landing 25/1/2004	Mars Exploration Rover-B (airbags)
Opportunity Rover	7/7/2003	2 S	354 E	Soft landing 25/1/2004	Carried by MER-B (+ other DO)
Beagle 2 Lander	4/6/2003	—	—	Crash? 25/12/2003	(Released from Mars Express orbiter)

(Continued)

Table 3 Lunar and planetary impact catalog[a] (continued)

Object[b]	Launch date[c]	Latitude	Longitude	Impact nature/date	Other debris/notes[d]
Planetary impacts: Jupiter					
Galileo Probe	18/10/1989	—	—	Crash 7/12/1995	Probably destroyed during entry
Galileo Orbiter	18/10/1989	—	—	Atmospheric entry Sept. 2003	Probably destroyed during entry
Planetary impacts: Saturn/Titan					
Cassini	15/10/1997	—	—	Orbiter (carried Huygens); in orbit July 2004	—
Huygens	15/10/1997	—	—	Soft landing on Titan 14/1/2005	—
Planetary impact: Asteroid 433 Eros					
NEAR-Shoemaker[m]	17/2/1996	—	—	Soft landing 14/02/2001	—
Planetary impact: Comet Tempel 1					
Deep impact	12/1/2005	—	—	Penetrator impact 4/7/2005	—

[a]Produced with the support of Nicholas Johnson, NASA.
[b]One or more objects left in lunar orbit may have impacted the surface by now.
[c]Dates are presented in United Kingdom format: day-month-year.
[d]Where other debris—released before impact/landing—is likely to have survived to the surface this is indicated.
[e]DO = debris object.
[f]DS = descent stage (of LM).
[g]AS = ascent stage (of LM).
[h]S-IVB = third stage of Saturn V.
[i]LRV = Lunar Roving Vehicle.
[j]Some Venus impact locations are incomplete.
[k]Spacecraft (s/c) bus probably destroyed during entry.
[l]MER = Mars Exploration Rover.
[m]NEAR = Near Earth Asteroid Rendezvous.

Fig. 47   First image of the Moon taken by a U.S. spacecraft, Ranger 7, before it added another crater to the lunar environment (taken on 31 July 1964 about 17 min before impacting the lunar surface: Mare Nubium is at centre left; large crater at centre right is the 108 km-diameter Alphonsus).

not insignificant objects—were crashed onto the Moon (along with the descent stage of Apollo 10). A seventh LM, that of the Apollo 13 mission, served famously as a lifeboat for the astronauts after an explosion disabled the CSM en route to the Moon; it was subsequently jettisoned and burned upon reentering the Earth's atmosphere.

Again, however, this amount of debris was almost insignificant compared with the five Saturn V rocket stages that were purposely crashed onto the lunar surface from the Apollo 13 mission onwards. The Saturn V comprised three stages, the third of which—the S-IVB stage—was used to boost the Apollo spacecraft on its lunar trajectory (Fig. 49). Once the Apollo 11 and 12 missions had placed seismometers on the Moon and found that there was little seismic activity, it was decided to use the S-IVB stage as a "thumper." The intention was that previously sited seismometers would detect the resulting reverberations, allowing lunar geologists to draw conclusions on the characteristics of the Moon's interior.

The S-IVB stage was about 18 m high, 6.6 m in diameter, and its dry mass (without propellant) was about 11 tonnes. Because the velocity required to escape the Earth's gravitational pull is 11.18 km/s (some 40,000 km/h),

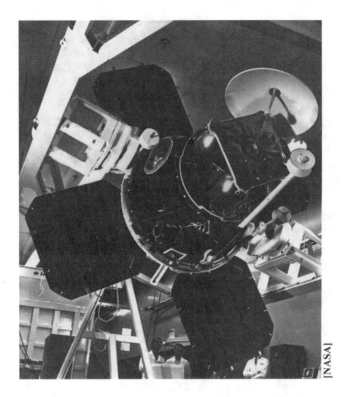

Fig. 48   Lunar Orbiter undergoes final preparations.

the stage would have significant kinetic energy by the time it impacted the Moon, even accounting for the fact that Earth's gravity would have reduced this somewhat on the journey. Little wonder, perhaps, that the Moon was observed to "ring like a gong" as a result of the S-IVB impacts.

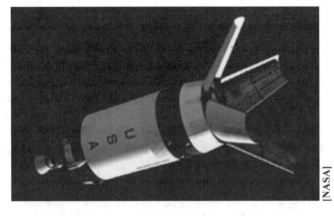

Fig. 49   A Saturn V (S-IVB) third stage following the release of its cargo, the Apollo lunar module.

An interesting appendix to the S-IVB story arose in September 2002 when a mysterious object in Earth orbit, at about twice the lunar distance, was discovered by an amateur astronomer.[6] Further investigations by professional astronomers at the Lunar and Planetary Laboratory of the University of Arizona concluded that the object—cataloged as J002E3—exhibited an optical spectrum consistent with white, titanium-oxide paint, the same as that used to cover the S-IVB stage.

By tracing the evolution of its orbit, NASA determined that J002E3 had been in solar orbit, in the vicinity of the Earth, until April 2002, when it is believed to have passed near the L_1 Lagrangian point (where the gravitational attraction of the Earth and Sun are in balance) and been captured by the Earth. The most likely candidate was believed to be the Apollo 12 S-IVB stage.

At first, there was speculation that the object could hit the Moon, but it was later determined that it would return to a (somewhat safer) solar orbit. It was interesting to note the apparent glee with which some lunar investigators rubbed their hands when it was thought that a lunar impactor had returned to the fold. There was talk of turning on the Apollo seismometers, for example, as if one more S-IVB impact would add significantly to lunar science. Perhaps when J002E3 returns in another 30 years, the voices raised will be those of concern rather than opportunity.

THE PROBLEM

The fact that many tonnes of spacecraft and rocket-body debris have been crashed onto the Moon is not widely known, even among the community of space professionals (that is, engineers, scientists, and other individuals engaged in space-related activity as a profession). Moreover, the majority of those who know about the impacts fail to recognize them as a problem. If it was not considered pollution in the 1960s, they ask, why should it be now? After all, the Moon's surface shows the result of meteor impacts that have occurred throughout geological time. What difference will a few more make?

Even the most hardened space environmentalist (should such an individual exist) might concede that a few spacecraft impacts on a body the size of the Moon have a relatively small effect, certainly compared with the megaton boulders that formed the impact craters we see from Earth. But simply ignoring them sets a dangerous precedent for the future of lunar exploration.

It is clear from the table of impacts that the practice of intentionally crashing spacecraft on the Moon did not cease with the Apollo impacts of the 1970s. Japan's Muses A/Hiten spacecraft crashed in 1993, and the impact of its sister craft Hagoromo is assumed, whereas as recently as 1999 NASA decided to target Lunar Prospector at the Moon's south pole.

In fact, when all of the spacecraft, propulsion units, and related debris that have landed or crashed onto the Moon are considered, the total number

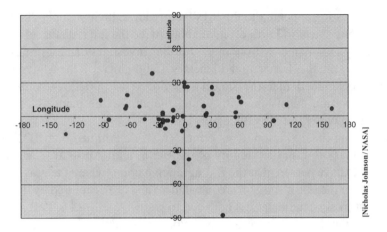

Fig. 50   Known landing and impact sites on the Moon.

exceeds 80,[7] and amounts to more than 100 tonnes of debris, the majority of which is concentrated near the lunar equator (see Fig. 50). Significantly, most of this debris was deposited in the course of little more than 10 years of lunar exploration.

Moreover, plans for the next couple of decades indicate an increased probability of further impacts (see Table 4). Europe has already launched its SMART-1 spacecraft to the Moon, and both India and China have announced plans to launch lunar orbiters. Japan, meanwhile, is preparing both landers and surface penetrators.

Interestingly, it is not only science spacecraft that might come close to the Moon, posing the risk of collision. For example, on Christmas Day 1997, the AsiaSat-3 communications satellite was stranded in a highly elliptical Earth orbit because of a malfunction of its Proton launch-vehicle's fourth stage. Following a transfer of ownership to Hughes Global Services (HGS), a subsidiary of the company that built the satellite, it was decided to engineer a novel lunar gravity-assist mission to transfer the satellite to an operational orbit.

From 10 April 1998, the renamed HGS-1 fired its apogee engine at 12 successive perigees (low points in its elliptical orbit), the last of which placed the satellite on a translunar trajectory. Thus, on 13 May, HGS-1 became the first commercial communications satellite to reach the Moon, passing within 6200 km of the surface before it headed back towards Earth.[8] Following a second lunar flyby to refine the final orbit, the satellite was maneuvered into an inclined geosynchronous orbit on 17 June.[9] At the time, Hughes believed that this was unlikely to be the last time a translunar injection maneuver would be used to recover a satellite following a launch-vehicle malfunction and was considering it as part of its operational portfolio.

Table 4   Planetary/interplanetary spacecraft currently in orbit or planned

Spacecraft	Comments[a]
Moon	
SMART-1 orbiter (ESA)	Launched 27 Sept. 2003; in orbit 17 Nov. 2004
Lunar Reconnaissance Orbiter (LRO)	NASA Orbiter; launch 2008?
SSTL Lunar Orbiter (UK)	Surrey Satellite Technology Ltd. minisatellite
Lunar-A (JAXA)	Lunar orbiter with penetrators
Selene (JAXA)	Polar orbiter + 2 subsatellites (ellipt'l orbits)
Selene-B (JAXA)	Lander and rover?
Chandrayaan-1 (ISRO)	Polar orbiter; launch 2007–8?
Chang'e (China)	Orbiter (2007), lander (2010), rover (2012), sample return (2020)
TrailBlazer (TransOrbital Inc)	Commercial orbiter; deorbit to impact
Icebreaker 1 (LunaCorp)	Lander and rover
Lunar Retriever 1 (Applied Space Resources)	Lander and sample return
Mercury	
Messenger (NASA)	Launched 3/8/04: in orbit by 2009 (Venus flybys 10/06 and 6/07)
BepiColombo (ESA)	Orbiter; launch mid-2012?
Venus	
Venus Express (ESA)	Orbiter (Launched 9 Nov. 2005)
Planet-C (JAXA)	Venus Climate Orbiter (2008)
Mars	
Viking 1 Orbiter (NASA)	In orbit since 1976
Viking 2 Orbiter (NASA)	In orbit since 1976
Mars Global Surveyor (NASA)	(Launched 1997; in orbit since 1997)
Mars Odyssey (NASA)	(Launched 7/4/2001; in orbit since 2001)
Mars Express orbiter (ESA)	In orbit since 2003
Planet-B/Nozomi (Japan)	Failed to enter Mars orbit in December 2003 (now in solar orbit)
Mars Reconnaissance Orbiter (MRO)	Launched Aug. 2005 (NASA)
Mars Scout 1—Phoenix (NASA)	Aug. 2007: lander targeted at N pole
Aurora: ExoMars orbiter (ESA)	2009? Descent module and rover
Mars Science Laboratory (MSL)	2011? Lander and rover (NASA)
Mars Scout 2 and 3 (NASA)	2011? Landers?
Aurora: Mars Sample Return (ESA)	2011?
NASA Mars sample return	2013?
Aurora: manned mission (ESA)	2030?

(**Continued**)

Table 4 Planetary/interplanetary spacecraft currently in orbit or planned (continued)

Spacecraft	Comments[a]
Pluto	
New Horizons (Pluto/Kuiper Belt)	Launch 2006; Jupiter flyby; Pluto by 2015?
Selected interplanetary spacecraft	
Deep Space 1 (NASA) to deep space	Launched 24/10/1998; science/technology demonstrator
Stardust (NASA) to Comet Wild 2	Launched 7/2/1999; rendezvous 2/1/04; sample return Jan. 2006
Genesis (NASA) to deep space	Launched 8/8/2001; solar-wind sample returned 2004
Contour (NASA) to Comet Nucleus Tour[b]	Launched 3/7/2002; destroyed in Earth orbit/SRM firing
Muses-C/Hayabusa (JAXA) to Asteroid Itokawa	Launched 9/5/2003; soft landing 11/2005; sample return 6/2007
Rosetta (ESA) to Comet 67P[c]	Launched 2/3/2004; orbit May 2014; carries Philae lander
Philae (ESA) to Comet 67P[c]	Launched 2/3/2004; soft landing Nov. 2014
Dawn (NASA) to Asteroids Vesta and Ceres	2006? Orbit Vesta 7/2010–7/11; orbit Ceres 8/2014–7/15

[a]Dates are presented in United Kingdom format: day-month-year. Unless otherwise noted, they are launch dates (future launch dates are always open to question).
[b]Encke and Schwassmann-Wachmann 3.
[c]Churyumov-Gerasimenko.

Meanwhile, as mission controllers discussed the potential use of lunar gravity for commercial satellites, an unassuming scientific spacecraft was using the same force to propel itself to Mars. On 4 July 1998, Japan's Planet-B (Nozomi) spacecraft was launched into a 400,000-km lunar parking orbit by an M-5 launch vehicle and, following two months of in-orbit testing, made the first of two gravitational slingshots to accelerate it towards a planned three-year mission in Mars orbit.

It is clear that lunar gravity is a useful resource for solar-system exploration. Unfortunately, it is also clear that this places the Moon at risk from unplanned spacecraft impacts should things go wrong.

Even if no further impacts occur in the future, the spacecraft and launch-vehicle stages that have already struck the Moon, producing craters and metallic debris on the surface, might one day present a safety hazard to lunar explorers and developers. It could also be argued, from a scientific viewpoint, that they have contaminated a formerly "pristine" planetary environment.

Indeed, the removal of defunct spacecraft by deorbiting has an arguably greater environmental impact for the Moon than it does for the Earth.

Because the Moon has no atmosphere, spacecraft and debris will impact the surface intact rather than burning up and disintegrating en route; and, of course, "ocean disposal" is not an option. Moreover, whereas the Earth's environment has a natural ability to repair itself, the lunar environment does not and will bear the scars of man's intervention for the foreseeable future, a future that can be measured in hundreds of thousands of years.

Should the tacit approval of the practice of impacting spacecraft on the lunar surface be continued? Would we be so keen to allow such impacts if there were inhabited bases on the Moon? One hopes not, but the situation with debris in low Earth orbit, and its damaging effect on manned spacecraft, is a far from encouraging model.

The record shows that the Moon is not the pristine body it once was, but what about lunar orbit?

LUNAR ORBITAL DEBRIS

The Soviet Union's Luna 10 spacecraft and its propulsion unit were the first man-made objects to enter lunar orbit, on 3 April 1966. They were followed over the next few years by other Luna spacecraft (some of which were separated from their propulsion units), the Lunar Orbiters (which were all removed from orbit), and a number of Explorer spacecraft. In addition, the Apollo 15 and 16 missions released small (approximately 36-kg) subsatellites into lunar orbit, both of which decayed and fell to the surface within two years (Fig. 51).

As of mid-2005, less than 50 man-made objects had been placed in lunar orbit. At least 34 have either crashed onto the Moon or been maneuvered out of lunar orbit (24 and 10, respectively).[4] Several of the remaining spacecraft and propulsion units are also believed to have reached the surface, but the complex lunar gravity field and the influence of the Earth make reliable predictions difficult. Although, ostensibly, the number of satellites in lunar orbit today is small, experience with Earth's orbital population gives cause for future concern.

First, there is the possibility of explosion and fragmentation as a result of the stored energy of residual propellant or battery constituents. This could produce clouds of debris that could damage other satellites and ultimately lead to the type of cascade problem discussed in the preceding chapter for low Earth orbit. Passivation of a spacecraft or propulsion unit at the end of its mission would significantly reduce its explosive potential and should, in the opinion of many experts, be mandatory.[10]

A second cause for concern is the solid rocket motors (SRMs), used, for example, by the Explorer 35 and Explorer 49 spacecraft for lunar orbit insertion. SRMs produce metallic particles as part of the combustion

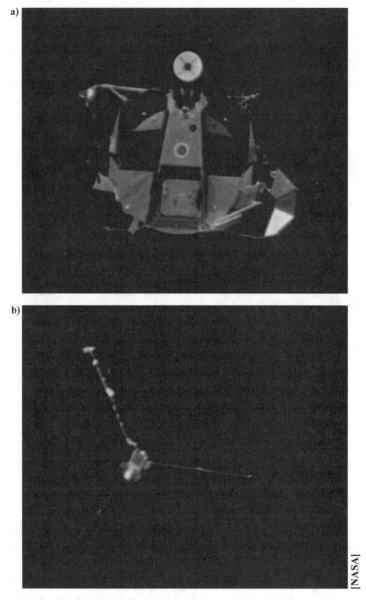

Fig. 51   Apollo lunar module ascent stage and sub-satellite, two examples of spacecraft that were deorbited onto the lunar surface.

process. Although it is too late to do anything about existing debris, the production of further similar material could be eliminated by using either liquid (chemical) propulsion systems or electric systems. However, if an SRM must be used, logic dictates that it should remain attached to the spacecraft rather than being jettisoned to form a separate debris object.

A third potential source of debris is represented by the spacecraft insulation materials and paint, which can become brittle in the harsh solar radiation and thermal environment and become detached. Bombardment by micrometeoroids can have much the same effect. Although the particles are likely to be very small, with short-lived orbits, experience in Earth orbit clearly indicates that vast numbers of these particles can accumulate.[10] The design of future lunar satellites and propulsion units should therefore incorporate appropriate countermeasures.

A final source of lunar orbital debris is a result of a mechanism not experienced in Earth orbit. Because of the Moon's relatively low surface gravity and the absence of an atmosphere, debris hitting the surface at great speeds can potentially be thrown onto high-altitude trajectories, or reorbited. As a result of a nearby perturbing force, in the guise of the Earth, these trajectories can then be modified into lunar orbits; put simply, the material goes up but it does not come back down. Observations made by the Galileo spacecraft in orbit around Jupiter have indicated that small pieces of natural debris produced in this manner might be in orbit around Ganymede, Callisto, and Europa.[11]

It is thus theoretically possible that the practice of deorbiting lunar satellites, and particularly the high-energy impacts of the Apollo S-IVB stages, produced not just surface damage, but also contributed to the orbital debris population. Without Moon-based radar systems similar to those of the Space Surveillance Network (SSN) on Earth, we have no way of knowing. And, of course, if the practice is continued in future, it could create new orbital debris. Although careful planning of impact trajectories and timing could reduce the potential for debris creation, the possibility of reorbiting increases the risks associated with the removal of objects from lunar orbit.

It has been suggested that damage to the lunar surface environment could be minimized by designating special regions along the lunar equator as "satellite dumping grounds," a policy that would also concentrate potentially valuable materials for subsequent salvage and use.[4] But if *reorbiting* of debris does in fact occur, this solution is far from ideal. Unfortunately for the Moon, deorbiting requires less energy than the alternative of boosting the spacecraft into solar orbit and is thus seen as more cost effective.

One of the fundamental problems of many lunar orbits is their inherent instability, resulting from the gravitational presence of the Earth. Among other things, this means that there is no lunar analog to geostationary orbit, whereby satellites would revolve above a given position on the lunar equator, and no analog to the graveyard orbit (see Box: Planetary Stationary Orbits). It would be possible to station a communications satellite, or group of satellites, at the Moon's L_2 Lagrangian point to provide communications for settlements on the far side, but this is analogous to only a single orbital position. Moreover, redundant or uncontrollable satellites would either have

PLANETARY STATIONARY ORBITS

A unique satellite orbit, known as geostationary orbit (GEO), exists around the Earth. It is a circular orbit in the same plane as the Earth's equator. Its radius is 42,164 km (26,200 miles), and its height above the Earth's surface is 35,786 km (22,237 miles).

A satellite in geostationary orbit has an orbital period equal to the Earth's sidereal period of rotation (23 hr 56 minutes 4 seconds) and therefore appears stationary with respect to the Earth (hence "geostationary"). This means that it can be allocated an orbital position related to the line of longitude above which it is stationed. GEO is the orbit used for most communications satellites, partly because it allows the use of small earth terminals, which do not need to track the satellite, thus decreasing the cost of the ground segment. In addition, a system of three equally spaced satellites can provide full coverage of the Earth, except for the polar regions where the elevation angle is very low.

Although a satellite in GEO is nominally stationary, a number of perturbing forces act to make the satellite's orbit unstable. The three main perturbation mechanisms are triaxiality (the nonspherical nature of the Earth), luni-solar gravity (the varying gravitational attraction of the Moon and Sun), and solar radiation pressure.

An analogous stationary orbit around Mars (areostationary orbit) would have a radius of about 20,400 km and an altitude of about 17,000 km. It too would suffer similar perturbing forces, although the effect of the gravitational attraction of the Martian moons, Phobos and Deimos, would have to be carefully calculated, because Phobos orbits at an altitude of about 9300 km above Mars (below areostationary orbit) and Deimos is above the orbit at about 23,400 km. Nevertheless, with an adequate stationkeeping system there is no reason why a stationary satellite in Mars orbit should not be viable.

The situation for the Earth's Moon is quite different, however, because a so-called selenostationary orbit would be highly unstable. The Moon makes a single rotation about its axis only once a month, which would mean that any stationary orbit around it would have the same period. Calculations show that its radius would be 87,678 km, placing it some 86,000 km above the Moon, which is about a quarter of the way to the Earth and well beyond the L_1 and L_2 Earth–Moon Lagrangian points, at about 65,000 km above the Moon.

The Lagrangian points are a set of points in space where gravitational forces are balanced so that a body at that point remains nominally stationary. They are useful for siting astronomical satellites (for example, the SOHO solar observatory, which is stationed at the L_1 Sun–Earth Lagrangian point).

What this means for the Moon, however, is that the L_1 and L_2 Earth–Moon Lagrangian points are the only stable positions for stationary satellites, and there is no lunar graveyard orbit analogous to that around the Earth.

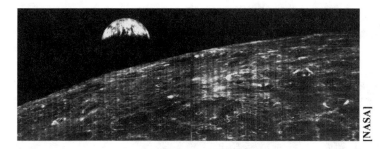

Fig. 52 Lunar orbit was relatively clear when Lunar Orbiter 1 took this first view of Earth from the Moon on 23 August 1966, but what about the future?

to be removed from the vicinity or allowed to wander among and beyond the operational spacecraft—hardly ideal. So, when a fully fledged satellite communications system in lunar orbit becomes necessary it will probably use satellites in low- or medium-altitude orbits. From an engineering point of view, this has many repercussions, not the least on the number of satellites required in what might become a constellation of lunar orbiters. It is difficult to ignore the inherent potential for both orbital and surface debris.

The future of lunar exploration is uncertain because it depends on significant funds being made available for lunar spacecraft and their operation. However, a number of missions are already being considered for launch in the next decade and, following the U.S. President's announcement of early 2004, many more are likely (as indicated in Table 4). For those that go ahead, it is clear that mission plans should include debris-mitigation measures. It is, however, less clear that such measures are actually being implemented.

Although the need for action to protect the Earth's orbital environment is now well recognized, it is important not to forget lunar orbit (Fig. 52). Without proper safeguards, an increase in lunar traffic could lead to the same type of debris problem in lunar orbit that we now have in low Earth orbit.

OTHER PLANETARY BODIES

Of course, the scientific exploration of the solar system does not end with the Moon. Spacecraft have successfully entered orbit around Mars, Venus, Jupiter, and Saturn, whereas others have soft landed on Mars, Venus, and a number of lesser bodies. The first in the latter category was NEAR-Shoemaker's landing on the asteroid Eros in 2001 (Fig. 53), while other missions include the landing of Muses-C/Hayabusa on asteroid Itokawa in 2005 and that planned for Rosetta's Philae lander on Comet Churyumov-Gerasimenko in 2014. A somewhat less benign encounter was the aim of

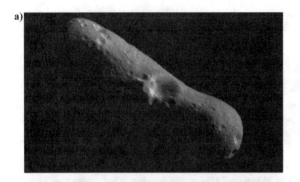

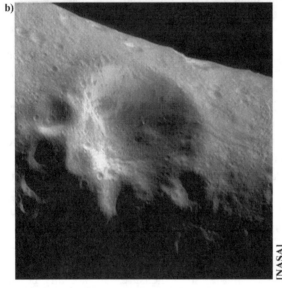

Fig. 53   The asteroid Eros is among the other solar system bodies visited by spacecraft (NEAR-Shoemaker landed there in 2001).

[NASA]

the aptly named Deep Impact, which fired a penetrator into Comet Tempel-1 on 4 July 2005 (Fig. 54).

As with the lunar landings, it is hard to characterize the successful landing missions, or even the penetrator probes, as polluting or degrading the celestial surfaces they are designed to investigate. In fact, these unmanned missions are likely to have a far less damaging effect than a manned mission, albeit with a more restricted scientific return. And one could hardly complain that the intentional injection of the Galileo entry probe into Jupiter's atmosphere would have a significant effect on that giant planet.

However, not all planetary spacecraft have been successful, and some have crashed, scattering debris on the respective planetary bodies (as detailed in Table 3). And although there is no conclusive evidence that any man-made debris resides in orbit around the planets—mainly because we have no way of tracking it—there is plenty of circumstantial evidence to suggest that it might, especially in the case of Mars.

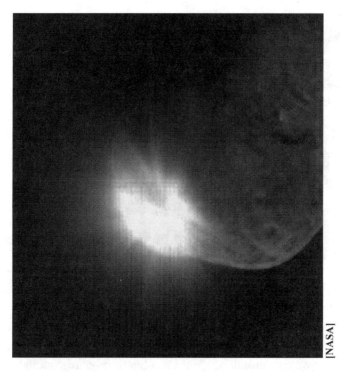

[NASA]

Fig. 54 America celebrated the 4th of July 2005 by smashing a 370 kg, washing-machine-sized impactor into Comet Tempel-1; this early image from the Deep Impact spacecraft shows the resulting explosion.

In the early years of Mars exploration, from 1960 onwards, so many Soviet spacecraft failed en route to the red planet that the Mariner 4 deputy spacecraft manager, John Casani, felt moved to devise a satirical explanation. He created the "Great Galactic Ghoul," a mythical creature that lurked midway between Earth and Mars ready to devour passing spacecraft.[12]

Unfortunately, it is so difficult to be sure what has actually happened to a spacecraft when its radio transmissions suddenly cease that it might just as well be the Ghoul at work! For example, even in more recent times, several spacecraft have been lost prior to entering Mars orbit or initiating a landing: Phobos 1 and 2 in 1989; Mars Observer in 1993; Mars Climate Orbiter and Mars Polar Lander in 1999.

The case of Mars Observer is particularly worrying because it might have exploded very close to Mars (Fig. 55). Contact was lost with the spacecraft just three days before it was due to enter orbit, following the pressurization of its propulsion system in advance of the deceleration burn. Because the space-craft's telemetry system had been deactivated and subsequent efforts to

Fig. 55   NASA's Mars Observer is believed to have exploded before
entering orbit around Mars in 1993.

communicate with it failed, it was not possible to establish conclusively what
happened. However, an independent investigation board concluded that the
most probable cause was a rupture in the fuel pressurization side of the propulsion system, which caused the spacecraft to spin uncontrollably.
Although a root cause of the event was believed to involve the inadvertent
mixing of nitrogen tetroxide (oxidizer) and monomethyl hydrazine (fuel),
which are hypergolic and ignite on contact, the board did not highlight an
explosion as a probable cause, but this is not the same as saying that an
explosion did not occur. The fact is, no one really knows what happened
to Mars Observer.

Apart from any technical problems that can occur with a spacecraft, one
must also consider the natural environment of interplanetary space, which,
to say the least, is not benign. Initially, it seems unlikely that something as
small as a spacecraft could be hit by debris—either natural or man-
made—in the immensity of space that exists between Earth and Mars, but
Mariner 4 proved otherwise, in 1964 and 1965, by recording 235 micrometeoroid impacts on its way to the planet.[12]

In 1969, when Mariner 7 lost its lock on the guide star Canopus a few days
from Mars, it was presumed that it was the result of an impact from a micrometeoroid.[13] Although control of the tumbling spacecraft was regained within
a few hours, its trajectory past Mars was found to have deviated by 130 km at
the point of closest approach. Later analysis suggested that the deviation was,
in fact, the result of a battery-cell explosion: electrolyte from a ruptured
battery cell is thought to have acted like a thruster, pushing the spacecraft

off course.[14] Alternatively, it is feasible that the battery exploded as a result of a micrometeoroid strike, but the truth is hard to ascertain.

Strangely, its seems that Mariner 7's sister spacecraft, Mariner 6, also lost lock on Canopus en route to Mars, when its camera scan platform was unlatched. Apparently, the operation released a bright piece of debris, which the star sensor immediately began to follow, instead of maintaining a lock on the target star.[14]

Whether one subscribes to engineering analyses or tales of the Galactic Ghoul, the important points are that, wherever they are, spacecraft can be damaged by debris and debris can be created by spacecraft. As with spacecraft in Earth orbit, the solutions lie in protection against the space environment, debris-mitigation techniques, and removal of the spacecraft from the sphere of conflict.

The deorbiting technique has been used on two notable occasions for two different planets: for the Venus radar mapping spacecraft Magellan in 1994 and for the Galileo Jupiter orbiter in 2003.

The philosophy of deorbiting these spacecraft once their missions had been completed was similar to that of deorbiting the Lunar Orbiters in the 1960s. On the one hand, once a spacecraft's useful maneuvering propellant has been depleted there is no way to control it, making it a hazard to other spacecraft; on the other, any residual propellant in the tanks might one day cause an explosion with equally damaging results. Although parts of Magellan could have survived reentry and crashed on the surface of Venus, it seems more likely that it was vaporized in the planet's thick atmosphere.

Certainly, there would be little left of Galileo, even assuming someone could descend through Jupiter's atmosphere to find it, so that in this case the decision to deorbit the spacecraft could provoke little dissent. The alternative was potentially more damaging because an uncontrolled Galileo could one day impact one of Jupiter's moons—specifically Europa, which might harbor some type of life-form. Thus it was that, following eight years of productive science in orbit, Galileo plunged into Jupiter's atmosphere on 21 September 2003.

Importantly for the current discussion, the fact that Magellan and Galileo were deorbited at all shows that some members of the space community believe in the need to protect at least some parts of the space environment.

PLANETARY PROTECTION

The protection of the planetary environments, which include the planets' moons, is generally led by the desires of planetary scientists to maintain the original "purity" of the environments they wish to investigate. The nub of planetary protection is encapsulated in the phrase "forward contamination," which can be defined as "the contamination of a planetary body other than the Earth, which may occur as a result of landing a spacecraft on that body,"[15] or indeed allowing it to crash there. The contamination of the terrestrial environment, which may occur as a result of returning samples from another planetary body, is known as back contamination.

The history of exploration on Earth can quote several examples of explorers taking diseases to isolated societies that had no immunity and wiping out their civilizations. Some worry that a similar thing could happen to indigenous life-forms, however simple, as a result of our exploration of the planetary bodies. Even more disturbing is the possibility that a sample returned from, say, Mars or Europa could contaminate the Earth. At present, this reads like many a science-fiction tale, but history has also shown that space exploration has a habit of turning science fiction into fact.

Surprising though it might seem, given the relative insensitivity of early space explorers towards the space environment, the first international effort at dealing with extraterrestrial contamination came as early as 1956, when the International Astronautical Federation (IAF) attempted to coordinate efforts at its seventh Congress in Rome.

Following the launch of Sputnik 1 in 1957, the United Nations General Assembly created the ad hoc Committee on the Peaceful Uses of Outer Space (COPUOS) and directed it to identify legal problems associated with the exploration of space[16]; this led, among other things, to the formation—by the IAF—of the International Institute of Space Law (IISL) in 1960. In addition to a Legal Subcommittee, COPUOS itself established a Scientific and Technical Subcommittee, which made the initial attempts to deal with contamination issues in 1958.[17]

Also in 1958, the International Council of Scientific Unions (ICSU) formed the Ad Hoc Committee on Contamination by Extraterrestrial Exploration (CETEX), which recommended that a code of conduct be established for space missions and research.[18,19] It also issued four primary directives (paraphrased here):

1) That there should be freedom of activity for experimentation and exploration of celestial bodies (within limitations based on planetary quarantine requirements);

2) That a coordinating committee should be informed of any proposed experiments, to allow it to study the probability of harm to the space environment and make recommendations to minimize it;

3) That experiments should only be carried out if they were likely to yield useful scientific data;

4) That no soft landings requiring the use of "large quantities of gas" should be made on the Moon prior to the completion of extensive studies of the Moon's atmosphere, and that no nuclear detonations should be made near its surface.

Apparently, mention of the Moon was intended to include the planets[19] and to include detonations *on* as well as "near" their surfaces.[20] One could quibble about the use of the word "gas" rather than something more general such as "propellant" and the vagueness of the word "extensive," but at least it was a start.

ICSU accepted the CETEX recommendations within the year and established the Committee on Space Research (COSPAR) to coordinate worldwide space research. One of COSPAR's first tasks arose in October 1961 when Project West Ford was announced (see Chapter 3). According to one author, "tremendous international controversy erupted concerning the potentially harmful effects" that the copper dipoles were expected to have, including interference with existing communications services.[19] As a result, COSPAR formed the Consultative Group on Potentially Harmful Effects of Space Experiments.

Interestingly, around the same time, science-fiction author Arthur C. Clarke was investigating the potential for forward contamination by means of a short story called "Before Eden."[21] In the story, astronauts exploring a high-altitude, polar region of Venus find a oxygen-producing plant that crawls across the surface. Before leaving the planet, they carefully bury their refuse in a plastic bag. Unfortunately, the creature devours the bag, along with the astronauts' bacteria and viruses, which kill it. The story represents a good example of how science fiction can provide the sort of "early warning" that academic papers cannot.

SPACECRAFT DECONTAMINATION

A growing realization that it might be necessary to protect the planetary bodies from contamination led, in 1964, to the publication by COSPAR of a set of planetary quarantine requirements (PQR), which were subsequently adopted by the United States. They were ostensibly more scientific than the CETEX code of conduct in that they were based on an attempt to establish the probability of contamination by a given mission. Appropriate precautions, such as decontamination or sterilization, were taken as a result of the statistical analysis,[20] a practice followed until 1982.[18]

Indeed, the possibility of contamination was considered from the very early days of the lunar exploration program. Research undertaken by the Rand Corporation had suggested that germs carried by a spacecraft from Earth could contaminate the lunar surface and invalidate any future searches for life. As a result, when America made its first attempt to send a spacecraft towards the Moon in August 1958, each section of the Pioneer spacecraft was sterilized with ultraviolet light before being assembled in a weather-tight room at the top of its launch gantry by technicians dressed in surgical gowns, gloves, and masks.

However, some questioned the need for sterilization. Dr. Harold Urey, one of the California Institute of Technology's most respected lunar scientists, thought that even if a square kilometer of the Moon were contaminated, some 30 million km^2 would remain untouched for future research. Besides, he argued, the harsh thermal and radiation environment would kill any microbes attached to the spacecraft.[22] Nevertheless, the early Pioneers and other planetary spacecraft were sterilized to avoid forward contamination.

Sterilization techniques were also used in the Ranger program, but because alcohol wipes and ethylene oxide gas could only reach parts of Ranger's internal mechanisms it was decided to heat the entire spacecraft to 125°C for 24 hours, despite evidence gathered in late 1961 that this would have "detrimental effects" on Ranger's electronic systems.[3] Unfortunately for Ranger 3, which was launched in January 1962, the concerns were well founded, and its central computer and sequencer failed from the effects of preflight heat sterilization. And although its successor, Ranger 4, earned the dubious honor of being the first spacecraft to impact the lunar far side, it was the practice of heat sterilization that stopped it taking any pictures of the event.

However, the sterilization policy was proved correct—at least to some— very early in the history of lunar exploration. In November 1969, as an interesting adjunct to the Apollo 12 mission, astronauts Charles Conrad and Alan Bean were able to visit the Surveyor 3 spacecraft, which had landed in April 1967 (Fig. 56). Having settled the lunar module a couple of hundred meters

[NASA]

Fig. 56 The Apollo 12 astronauts visited Surveyor 3 in November 1969; an accurate landing allowed them to return parts of the probe to Earth.

from the Surveyor, they walked to the spacecraft, removed selected hardware, including a camera, and returned it to Earth for analysis. Amazingly, when the camera was dismantled back in the lab, technicians reportedly discovered *streptococcus mitis* bacteria on a sample of foam inside the camera housing.[23,24] According to one author, the surface wrappings of certain electrical cables had been "deliberately contaminated with several thousand Bacillus Subtilis spores."[25] Whatever the biological identity of the stowaways, and however they got there, they had seemingly not only survived the journey to the Moon, but also two years and seven months on the lunar surface. The requirement for sterilization had apparently been demonstrated.

However, as with much of science, it is difficult to obtain definitive proof. Other authors have suggested that the contamination might have been introduced in NASA's Lunar Receiving Laboratory (LRL). It is reported, for example, that no other components of the Apollo 12 camera contained bacteria, and none were found in the test camera that remained on Earth.[26] The truth might never be known, but the streptococcus controversy illustrates the potential for confusion in analyzing planetary sample-return payloads and the need to take both forward and back contamination seriously.

In general, spacecraft intended to land on planetary bodies are designed under the assumption that they will analyze chemical constituents native to that body and not contamination from Earth. It was for this reason that each of the Viking spacecraft, which landed on Mars in 1976, was sterilized before launch. By then, heat sterilization techniques had improved, and electronic components were more rugged, so the entire spacecraft could be placed in an oven, heated to 113°C, and cooked for about 40 hours. To minimize the oxidation of spacecraft components, a 97% pure atmosphere of nitrogen gas was circulated through the chamber during the process, and to ensure that everything worked afterwards the landers' systems were tested before encapsulation in the launch vehicle.[27] In addition, the spacecraft were each equipped with an extensible robot arm, which enabled soil samples to be collected some distance from the lander, thereby avoiding significant contamination from the vehicle's hydrazine descent engines (see Fig. 57).

COSPAR's original planetary quarantine requirements continued their evolution throughout the 1970s, presenting what at first sight appeared to be a complicated and comprehensive formula (see Box: Probability of Contamination Formula). However, it was criticized by the Space Science Board of the U.S. National Research Council/National Academy of Sciences (NAS) because it required users to assign probabilities to phenomena that could not be verified with any precision. Moreover, it could produce inconsistent results with only slightly different values, most of which were estimates anyway.

Fig. 57   A technician inspects the soil sampler on the Viking lander's robot arm. The spacecraft was sterilized before launch as part of a planetary protection program.

PROBABILITY OF CONTAMINATION FORMULA[19]

NASA's planetary quarantine program of the 1970s devised a formula to determine the probability of contamination[25]:

$$Pc = mi(0) \ P(vt) \ P(uv) \ P(a) \ P(sa) \ P(r) \ P(g)$$

where

Pc is the probability of contamination;
mi(0) is the initial microbial burden at launch after decontamination;
P(vt) is the probability of surviving space vacuum and temperature;
P(uv) is the probability of surviving ultraviolet radiation;
P(a) is the probability of arriving at the target planet;
P(sa) is the probability of surviving atmospheric reentry;
P(r) is the probability of releasing an organism in a viable state; and
P(g) is the probability of growth of that organism in the space environment.

Indeed, originally, the PQR were based on calculations intended to cover *all* potential interplanetary missions to a particular body in an initial period of 10 years, such that the probability of contamination by a single viable organism was to be less than 1 in 10,000 for a lander and less than 1 in 300,000 for a flyby mission. Later, the probability of contamination for landers was reduced to 1 in 1000. To complicate matters, the probabilities were to apply, not to any individual mission, but in the aggregate for all missions conducted during that period, an aggregate apportioned between the major spacefaring nations. As a result, almost half of the U.S. allowance was allocated to the Viking missions alone (Tennen, L., personal communication, 11 June 2004).

In 1978, the quarantine requirements were substantially revised to take into account work done by the Space Science Board, which showed that the probability of contamination could be reduced by an order of magnitude.[28] Ostensibly, this eliminated the need for *any* decontamination procedures on spacecraft destined for the outer planets, Jupiter, Saturn, Uranus, and Neptune.

Some observers at the time complained that the accepted value for the probability that an organism would grow in the space environment was so low as to exclude *most* missions from the decontamination requirement.[29] Their view was supported by the further relaxation that occurred in the 1980s to the extent that, as far as NASA was concerned, planetary protection measures were no longer the norm, but could be determined on a case-by-case basis.

It is unfortunate, and not especially prudent, that the relaxation of policy occurred during a period when no new scientific data were being gathered concerning possible life on Mars, or any other planetary body. Significantly, no new data were available to build upon the inconclusive results from the Viking landers. It seems fair to conclude that the limitation of decontamination measures was a result of financial concerns rather than an application of the scientific method.

Indeed, in retrospect it appears that the PQR formula was a cul-de-sac on the roadmap for planetary protection. Over time, work by COSPAR and space agencies such as NASA has led to the replacement of the formula by a table, which eliminates the need to assign and calculate probability values based on uncertain data. It divides potential missions into five categories and allocates a degree of environmental concern to various target bodies, the higher-numbered categories being the most "at risk" (see Table 5 and Appendix A). Interestingly, in terms of nomenclature, what were originally known as planetary quarantine requirements became a Planetary Protection Policy,* and the field entered a new phase (Sterns, P., personal communication, 26 April 2004).

---

*COSPAR's Planetary Protection Policy website is http://www.cosparhq.org/scistr/PPPolicy.htm; NASA's Planetary Protection Policy website is http://planetaryprotection.nasa.gov/pp/missions/index.htm.

Table 5   COSPAR planetary protection policy: summary of proposed categories for solar system bodies and types of mission[18,a]

Category	Mission type	Target bodies (others to be determined)	Degree of concern/ requirements
I	Any but Earth return	Venus Undifferentiated, metamorphosed asteroids	None
II	Any but Earth return	Jupiter, Saturn, Uranus, Neptune, Pluto/Charon Comets Carbonaceous chondrite asteroids Kuiper Belt objects	Record of planned impact probability and contamination control measures
III	Flyby, some orbiters (no direct contact)	Mars Europa	Limit on impact probability Passive bioload control
IV	Lander, probe some orbiters (direct contact)	Mars Europa	Limit on probability of non-nominal impact (e.g., hard landing) Limit on bioload (active control)
V	Earth return	Mars Europa	No impact on Earth or Moon Returned hardware sterile Containment of any sample

[a] A NASA protection policy of 1999, which has similar categories, expanded the description as follows[28]:

Category I (which included Sun, Moon, and Mercury): "Not of direct interest for understanding the process of chemical evolution. No protection of such planets is warranted and no requirements are imposed."

Category II (which included the other planets except Mars and Earth, and outer planet satellites except Europa): "Of significant interest relative to the process of chemical evolution but only a remote chance that contamination by spacecraft could jeopardize future exploration."

Category III and IV (which included Mars and Europa): "Of significant interest relative to the process of chemical evolution and/or the origin of life or for which scientific opinion provides a significant chance of contamination which could jeopardize a future biological experiment."

Category V included "any solar system body."

Planetary protection policies were revisited in 1994 and 1999, and the probabilities of contamination continued to be revised, specifically in relation to Mars and the prospect of discovering indigenous life there. NASA generally adhered to a policy of sterilization (meaning "the complete elimination of organic life") for Mars landers and decontamination (meaning "a specific reduction of the microbial burden") for flybys and orbital missions.[30] However, spacecraft not actually carrying instruments designed to detect life were subject to substantially less stringent decontamination techniques.

The application of the category system is perhaps best illustrated by considering a number of specific cases. The Galileo Jupiter orbiter and its atmospheric entry probe, for example, were assigned a category II, specifically because of the low probability that any bio-organisms they carried to Jupiter would survive there. NASA decided, therefore, that decontamination of the spacecraft was unnecessary.[30] Mars Observer, on the other hand, was given a category III—even though it was intended as an orbiter rather than a lander—mainly because bacteria were thought more likely to grow on Mars.[31]

Naturally, a higher category still is awarded to spacecraft intended to land on Mars, such as Mars Pathfinder, which was launched on 3 December 1996 and on 4 July 1997 delivered a small rover—Sojourner—to the surface of the planet (see Fig. 58). Because the Pathfinder Lander and Sojourner Rover were intended to make contact with the surface of Mars, this mission was categorized Category IV-A for planetary protection purposes and was subject to appropriate requirements for preventing forward contamination. However, this does not mean that the bacterial count or "bioload" was zero. The requirement was only that it should be reduced to less than 300 bacterial spores per square meter and no more than 300,000 in total. At the time of launch, planetary protection personnel estimated that there were less than 24,000 spores on the craft.

An identical classification was applied to the Mars Exploration Rovers, Spirit and Opportunity, which were launched in June and July 2003, respectively, and landed on Mars in January 2004. Those spacecraft parts with large surface areas that could tolerate high temperatures (for example, the airbags used during landing) were subjected to dry heating to reduce the bioload. Electronics boxes and other sensitive areas were protected from external biological contamination using high-efficiency particulate arrestor (HEPA) filters, and alcohol wipes were used during assembly to maintain the cleanliness of each rover. Although sampling prior to encapsulation in the aeroshell showed that the total spore count on both rovers was well

Fig. 58   Mars Pathfinder and the Sojourner rover on Mars. The mission was awarded planetary protection category IV-A to provide protection for the Martian environment.

below the allowable level, the spore count for each spacecraft has been estimated to be 200,000.

Although the European Space Agency has conducted far fewer planetary
missions than NASA, it too follows COSPAR's Planetary Protection
Policy. For example, ESA's Mars Express orbiter was initially targeted off
Mars for planetary protection reasons (in case it should become uncontrollable
and crash onto the planet). It was then retargeted to deliver the Beagle 2 lander
to the surface before being retargeted again to place Mars Express in orbit
(Fig. 59).

Despite these attempts to protect planetary environments from contamination, it is clear from the preceding information that the simple
existence of planetary protection guidelines is not enough to ensure
anything like full protection. As pointed out by ESA lawyer Ulrike
Bohlmann, COSPAR is a nongovernment organization "not endowed with

Fig. 59   Three different categories of spacecraft for planetary protection purposes: a)
Mars Express orbiter; b) Mars Exploration Rover; c) Mars sample return concept.

institutionalized authority," and its recommendations are not legally binding—they simply represent "a moral kind of obligation." Moreover, although NASA and ESA have adopted the COSPAR guidelines, they are not binding internationally.[32]

A further complication is the lack of continuity in the field of planetary exploration, which tends to make the application of special guidelines an exceptional event, rather than the norm. There are simply too few space missions to achieve a situation where planetary protection is considered "part of the job," in the same way as reliability analysis and thermal-vacuum testing, for example. The two-decade hiatus in Mars surface exploration that followed the Viking missions was finally broken by the Mars Pathfinder mission in 1997. A similar gap in lunar exploration separated the Luna 24 mission in 1976 and the crash of Muses-A/Hiten in 1993. In the career cycles of engineers and managers, this was long enough for those experienced in planetary protection to have moved on, or retired, and been replaced by those with no direct experience of the issues.

Certainly, by the time the plan was made to crash Lunar Prospector (Fig. 60) onto the Moon's south pole, on 31 July 1999, the risks and costs associated with spacecraft sterilization seem to have outweighed the benefits in the eyes of spacecraft designers and their managers. The spacecraft was not sterilized or subject to active decontamination measures.[28] Considering the reasoning behind the impact—ostensibly an attempt to produce a plume of water molecules from suspected polar ice deposits—the decision is seen as particularly ill advised.

Moreover, as a public relations adjunct, Lunar Prospector also delivered a capsule of cremated human remains into that pristine polar environment. Although it might seem fitting, from an emotional point of view, that the cremated remains of highly regarded lunar geologist Eugene Shoemaker should finally come to rest on a planetary body he spent his lifetime observing, it is hardly consistent with the search for potable water. Some would go so far as to say that it has set "a terrible precedent."[28] The sad fact is that the mission was not a violation of the contemporary planetary protection policy, because it exempted lunar missions, condemning them as not "biologically interesting."[28]

By contrast, Lunar Prospector's predecessor Clementine, the first space-craft to suggest the existence of lunar water ice, was maneuvered out of lunar orbit in 1994, having spent less than three months in orbit.[4] Either different protection policies were followed for different spacecraft or, more likely, any existing policy was overridden by respective mission plans. The impact of Lunar Prospector served only to convince those already critical of planetary protection policies that they were in urgent need of an overhaul.

Fig. 60   Lunar Prospector in a clean room environment prior to
launch; it ended its mission by crashing into the pristine
environment of the Moon's south pole in the hope that its
impact would produce a plume of water molecules from
suspected polar ice deposits—no visible plume was detected.

POLICY REFINEMENT

In 2002, the refinement of COSPAR policy was addressed at a joint
COSPAR/IAU (International Astronomical Union) workshop on planetary
protection, which considered among other things the potential forward contamination of Europa and Mars (Fig. 61).

Protection for Europa is based on evidence that an ocean of liquid water
might exist beneath the icy crust. Noting that terrestrial life has been
shown to adapt to "extreme conditions, such as heat, cold, pressure, salinity,

a)

b)

[NASA]

Fig. 61 Protection for Mars and the Jovian moon Europa is uppermost in the planetary protection regime. Concepts for a) a "cryobot" mission to probe Martian polar ice and b) a "hydrobot" to explore an ice-covered ocean on Europa. A cryobot would move by melting the ice ahead of it and allowing it to refreeze behind.

acidity, dryness, and high radiation levels as well as combinations of them," the workshop concluded that an organism from Earth could "colonize the entire subsurface via the ocean connection." Put simply, a single impact of a contaminated spacecraft could affect the entire moon.

The workshop's recommendations included applying planetary protection requirements to future orbiters as well as landers, because an orbiter would have a significant probability of impacting the surface eventually.

The mere existence of so-called "extremophiles" on Earth encourages the belief that life could exist in what we humans might consider less than comfortable environments elsewhere in the solar system. Specifically, the notion that life might exist on Mars, despite Viking's negative results, was encouraged by research conducted in Chile's Atacama Desert in 2003.[33] NASA scientists concluded that if Viking had landed in the Atacama, its instruments—insufficiently sensitive to detect the low levels of organic material found there—would have produced the same result, with the implication that there was no life on Earth.

The possibility of contaminating Mars was highlighted in 2003 by experience with Japan's Mars probe, Nozomi, which developed problems with both its power and propulsion systems en route to Mars. Some scientists worried that if Nozomi could not be adequately controlled, it might impact Mars rather than enter orbit, while others suggested that if it could not operate its science instruments because of a lack of power, then an attempt at orbital insertion should be abandoned.[34] As it turned out, the mission was curtailed after an attempt to fire its engines failed, and the probe flew past Mars into an orbit around the Sun.[35]

Despite advances in planetary protection policies, there is a tendency to restrict consideration of a contamination event to a single, isolated mission. To do so would be a mistake. It is possible, for example, that a contaminated probe could take a terrestrial organism, microbe, or virus to another planetary body and deposit it there (and because current standards allow for a prelaunch contamination level of 300 bacterial spores per square meter for Mars landers,[16] this seems perfectly possible). A later spacecraft carrying a more sensitive instrument might then detect the contamination and identify it as an *extraterrestrial* life-form. The whole premise of exobiology would be at risk.

If this seems like science fiction, consider the Surveyor camera: if the Apollo 12 astronauts had not visited the spacecraft, the bacteria might have remained detectable for decades, and future lunar explorers, unaware of the source, might well have concluded that they were extraterrestrial in origin. This is analogous to failings in biological science exposed by Nobel physicist Richard Feynman when he investigated the alleged learning abilities of rats in maze experiments. He showed that experimenters had failed to decontaminate the maze sufficiently after each experiment, and, rather than learning the route, the rats had simply followed their noses! So the question is: when scientists discover the common cold virus on Mars, how will they be sure that it did not originate in the Jet Propulsion Laboratory? Furthermore, now that detection techniques are so sensitive, we run the risk of discovering DNA on a planetary body and assuming it to be extraterrestrial—it would take only a tiny hair or a fragment of dandruff deposited in a spacecraft cleanroom to lay a trail of innocent deception.

The COSPAR/IAU workshop also discussed the potential for back contamination resulting from sample-return missions, recommending that either entire spacecraft or "the subsystems involved in the acquisition, delivery, and analysis of samples" should be sterilized to "Viking post-sterilization biological burden levels."[18] So it seems that opinion on sterilization, if only for sample-return missions, has come full circle, a development that should be welcomed.

Of course, this will involve an increase in mission costs, but with science spacecraft typically costing tens or even hundreds of millions of dollars, a few more million for a sterilization program seems money well spent. There is, quite rightly, a move to limit the costs of these programs, but if costs must be limited, surely it is preferable to remove an instrument rather than remove the possibility of detecting indigenous life-forms.

Moreover, as the workshop report points out, there is another, practical reason to avoid contamination that might lead to a "false positive," in that such a result "would inhibit distribution of the sample from containment and would lead to unnecessary rigor in the requirements for all later Mars missions."[18] This suggests that spacecraft sterilization, in common with most preventative measures in life, will pay for itself in the long run.

PROTECTION RATIONALE

But why is all this effort regarding planetary protection so important when some would have us believe that a form of interplanetary cross-pollination has been going on for millennia? The evidence for this comes from meteorites found on Earth but shown to have the constituents or isotope ratios of other planetary bodies.

The most famous example is the so-called Martian meteorite, found in the Allan Hills region of Antarctica and designated ALH84001. In 1996, NASA announced that it appeared to contain fossilized life forms, thereby suggesting that there was once life on Mars (Fig. 62). Less well known is the lunar meteorite designated ALH81005, which, among others, is believed to have reached Antarctica from the Moon.

We have also been told that living organisms could have been brought to Earth by meteorites, and that flu and other pandemics might have been delivered by comets.[36,37] If true, these are natural events over which mankind has no control. However, knowingly delivering our bugs and

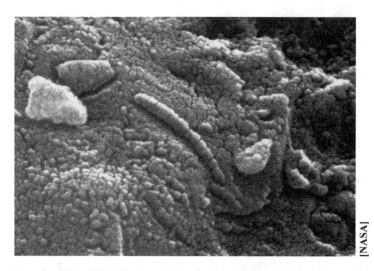

[NASA]

Fig. 62   The ALH84001 Martian meteorite, believed by some to contain the fossilized evidence of former life-forms.

diseases to another planetary body is quite another matter and should be avoided. Apart from anything else, one never knows when we might unwittingly bring back a radiation-mutated version of a terrestrial pathogenic virus, to which we have no immunity.

A less speculative rationale for planetary protection is biological science's search for the "second data point,"[38] a second place in the universe where life exists. The ramifications of finding and confirming the existence of indigenous extraterrestrial life are so fundamental—in both scientific and human terms—that any chance of doing so must be ensured. Indeed, the National Academy of Sciences Space Science Board considers the protection of Mars from terrestrial contamination to be "profoundly important."[39] It remains to be seen whether this view is shared among the space science community at large.

Naturally enough, given its importance to science, the debate on planetary protection has concentrated on biological contamination, but it is important not to lose sight of its nonbiological sibling in the process. The uncertainty produced by possible spacecraft hardware contamination can be time wasting and misleading in the pursuit of scientific truth, as shown by some thread-like features discovered in the Martian soil at the Opportunity rover's landing site in April 2004.[40] Images taken by the rover's microscopic imager (MI) showed millimeter- to centimeter-sized threads, which could not conclusively be labeled as "indigenous to the planet." It was thought likely that the fibers were part of the rover's airbag landing system rather than some Martian life-form, but the surmise could only be proved if similar, supposedly indigenous structures were found well beyond the landing site (which even then would be inconclusive because they could have been carried there by the wind or by the rover itself). This uncertainty simply underlines the need to take forward contamination seriously.

Like the planets themselves, planetary protection policy is a moving target. If it is to be at all effective it must respond to, or even pre-empt, plans for actual planetary missions.

FUTURE EXPLORATION

The dispatch of unmanned spacecraft to the planets and their moons is, however, only the first step in our exploration of the solar system. Just as astronauts followed satellites into Earth orbit and, later, to the Moon, so will they one day venture towards the planets. Indeed, the Moon has often been mentioned as an ideal environment for an astronomical observatory because of its limited atmosphere, surface stability, lack of erosion or wind loading, and the availability of natural radiation shielding.[41]

The problem, as our limited experience of manned lunar exploration has shown, is that explorers take their equipment, their infrastructure, and their habits with them. And if no binding, anticontamination or nonpollution agreements are in place, the explorers could leave an indelible and irreversible imprint on the environment they are attempting to study.

The history of exploration of terrestrial wilderness areas provides a useful analogy to this new phase of exploration. In the early days of Himalayan exploration, for example, Earth's highest mountain was considered a challenge for human endurance, an obstacle to conquer, and even—as a result of failed attempts—a killer. Attempting to reach the summit of Mount Everest was considered the laudable occupation of true heroes.

The fact that the slopes of the mountain have been littered with empty oxygen cylinders and other detritus as a result of the many attempts was accepted, for a long time, as a necessary prerequisite for conquest. It was also the case that, although ordinary hill walkers were encouraged not to despoil the natural environment with their picnic remnants, professional explorers risking life and limb in their endeavors were given more leeway. However, now that thousands of people have been on Everest, we are less inclined to make such allowances, and a program to clean up the mountain has been initiated, returning the discarded oxygen bottles and other rubbish to base camp and beyond. Interestingly, some of the earlier bottles have ended up in museums.

The next group of lunar explorers must learn from this, not least because of the Moon's inability to repair itself or even hide the evidence. After all, the Moon has no snow to cover oxygen cylinders and no glaciers to hide abandoned equipment, crashed vehicles, and dead bodies. It will be necessary to adopt the philosophy of the enlightened weekend hill walker and "take all litter home."

As permanent scientific bases are established on the Moon, the need for rules and regulations will become ever more important, because the potential for environmental damage will increase. The analogy here is provided by the science bases of Antarctica, the largest of which resembles a small town. Although guidelines exist for Antarctic "residents," where there are people there is pollution. It seems to matter little that the majority of those people are scientists tasked with studying the environment; they still need the occasional reminder.

The legacy of several decades of study and exploration was noted by polar explorer Ranulph Fiennes in his book about his journey across Antarctica: "My lasting memory of our brief visit to the Pole was disappointment that the Americans had let such a sacrosanct place deteriorate into an ugly, cluttered dump," he writes.[42]

According to Fiennes, the U.S. Base at MacMurdo Sound initiated a $30 million clean up in 1989, in the course of which "60,000 pounds of waste, thousands of drums of oil and solvents, old transformers containing PCBs, asbestos, explosives and chemicals were removed."[42] It seems clear that similar clean-up campaigns would be required for extraterrestrial bases and that they could cost proportionately more because practically everything we do costs more in space.

Unfortunately, our limited experience in manned lunar exploration indicates that we are likely to repeat the early Antarctic experience: in addition to leaving radioactive power sources and chemical batteries on the Moon, the Apollo crews dumped their three-days' rubbish and waste before departure, to reduce the liftoff mass of the lunar module. Of course, most reasonable people would subscribe to the "early days" excuse for this behavior, but this cannot be used forever. Cleaning up the Earth is a matter of balance, and so it will be for the Moon and any other planetary bodies mankind wishes to visit and study (Fig. 63).

U.S. President George Bush, unwittingly, moved the issue up the agenda in January 2004, when he announced America's intention to return astronauts to the Moon. Moreover, he added, "With the experience and knowledge gained on the Moon, we will then be ready to take the next steps of space exploration: human missions to Mars and to worlds beyond."[43]

The 2002 COSPAR/IAU Workshop on Planetary Protection considered what many—apparently including Bush—believe to be the inevitable follow-up to unmanned missions to Mars: exploratory visits by astronauts, permanent manned bases, and eventual settlements. However, its participants warned that manned missions should be preceded by unmanned, including sample-return, missions that would provide scientific information about the planet and identify potential hazards to future visitors. A specific

[NASA]

Fig. 63   Will future lunar bases be like existing Antarctic settlements, or will more care be taken of the lunar environment?

line of research concerned analysis of the Martian dusts "to see if they could be certified as 'safe' before any humans arrived," and it was recommended that systems should be designed to "break the chain of contact" with the Martian environment to avoid back contamination to Earth.[18]

Another important topic was the potential forward contamination of Mars by the astronauts, the suggestions being that special regions should be afforded protection and that all activities and operations should protect future biological examination of the planet.

In fact, there is a precedent for this in the Apollo program, which made at least some attempt to reduce lunar contamination. For example, a bacterial filter system incorporated into the lunar module was designed to prevent contamination of the local environment when the cabin atmosphere was released prior to an EVA.[44] NASA also adopted, as official policy, "aseptic subsurface drilling, decontamination and contained storage of waste materials, and biological and organic material inventory requirements."[45] Like all good surgeons, they sterilized their tools, collected their swabs, and made sure nothing harmful came into contact with the patient.

As far as protecting Mars from astronauts is concerned, many observers are pessimistic, even cynical, believing that once a manned mission is underway, any planetary protection policy will become academic.[38,46,47] Even with the best will in the world, accidents happen, pollution occurs, and environments are damaged. The fact that Mars experiences planet-wide dust storms on a fairly regular basis shows that isolating the effects of a propellant leak, for example, would be nigh on impossible.

Until late 2003, it was fairly clear who the main players in any future manned exploration of the solar system would be, but China's launch of its first astronaut in November of that year added another variable to the equation. The nation's plans for lunar exploration were reported by China's *People's Daily* newspaper, which quoted Luan Enjie, the vice minister of the Commission of Science, Technology and Industry for National Defence (COSTIND) and director of the China National Aerospace Administration (CNAA), as saying: "[with] the world programme of returning to the Moon not yet in full swing, we must seize the opportunity and start China's lunar exploration project as quickly as possible, to ensure that China has a niche in the international lunar exploration activity."[48]

Although China's first missions would be unmanned, it was clear from Luan's statement that China's ambitions include manned missions. He even added that building permanent bases on the Moon was "a vitally important first step in human development of outer space resources and expansion of living space."

Although we are used to hearing such sentiments expressed by media-savvy westerners, and treat them as part of the background flux of pro-space proselytizing, it is unusual to hear this from China. The fact that another nation has the capability and the desire to explore our nearest celestial neighbor makes international agreement on protection of the space environment that much more urgent.

There is, moreover, life beyond scientific exploration and study: as in Earth orbit, exploration will be followed by commercial and industrial development, bringing with it, inevitably, an even greater potential for damage to the space environment.

SPACE TOURISM

So far, the commercial development of the solar system has been limited to communications satellites in geostationary orbit and imaging satellites in sunsynchronous orbit. There are indications that this is changing, however. The construction of the International Space Station (ISS) in low Earth orbit and the desire on the part of the Russian partner to make money from the surplus third seat in its Soyuz resupply capsules has produced a nascent space tourism industry. The first person to take advantage of this was the American billionaire Dennis Tito, who paid a reported $20 million for a week's trip to the ISS. The second was South African businessman Mark Shuttleworth, the third was scientist Greg Olsen, and others have undergone evaluation and training for possible future flights.

Of course, very few people are willing or able to afford such a trip, but a precedent has been set, and the possibilities have been demonstrated. Unfortunately, space tourism is technology-limited, in that vehicles that would make it sufficiently cheap and safe for less wealthy adventure tourists to visit LEO do not exist. This is no reason to dismiss it, however. There was a time, still fresh in the memory of many alive today, when lunar exploration vehicles and space stations themselves did not exist.

Although orbital tourism might be some way off, suborbital tourism could be closer than we think. On 27 September 2004, Virgin's chief executive Richard Branson announced that he was entering the space tourism business with Virgin Galactic, using a vehicle based on the X-Prize-winning Space-ShipOne. Virgin sponsored the attempt to win the Ansari X-Prize as part of its overall deal to license the technology of SpaceShipOne for the purposes of space tourism. Branson said he expected his first spacecraft, the VSS Enterprise, to enter service by 2007, adding that "one day—hopefully in our life times—we would like to see a Virgin hotel in space."

The growing interest in spaceflight as a form of adventure tourism implies that, as long as human space exploration and development continues, space tourism is inevitable. And once tourism is established in LEO (hopefully with an infrastructure that protects the future use of the orbital resource), it follows that someone will want to go further: to the Moon. Apollo showed that it was technically possible. In fact, once a spacecraft is in LEO most of the hard work has been done, and boosting a spacecraft to the Moon is a relatively simple matter given today's technology. Moreover, it is only a three-day journey and far less strenuous than a hike to the foothills of the Himalayas, which is what some people consider a holiday.

Immediately one considers the possibility of lunar tourism, one has to ask "where would the tourists want to go?" It is a pretty safe bet that close to the top of the list would be the Apollo 11 landing site, the place where men first set foot on a celestial body other than the Earth. How many lunar tourists would give their eye teeth to gaze on that famous boot print photographed in 1969 by Buzz Aldrin and ever thereafter featured in books, magazines, and TV documentaries around the world? The temptation to place one's own boot in that finely delineated depression in the dust (Fig. 64) would, for some, be almost overwhelming. Imagine, by analogy, the thrill of the Everest climber if it were possible to find Edmund Hillary's boot prints in the snow!

But there lies the difference: footprints on Earth are transitory and ephemeral (unless of course they are those of long-dead dinosaurs preserved in stream beds); footprints on the Moon could remain unchanged for millennia. And from a cultural perspective at least, it is reasonable to assume that Aldrin's boot print ranks higher than that of any dinosaur. The logical conclusion, therefore, is that it should be preserved, in the same way that we preserve historical and cultural artifacts here on Earth.

If the Apollo landing sites, and others, are to be valued as historic sites and possibly become revenue-earning tourist attractions, they must be preserved

[NASA]

Fig. 64   The ultimate iconic image of the 20th century: how long will Aldrin's boot print survive once the lunar tourists arrive?

from accidental damage or trophy hunters. Perhaps a future U.S. Congress will vote for the Apollo exploration areas to be designated as the equivalent of national historic sites. Indeed, because the sites are not part of U.S. national territory, perhaps an international body such as the United Nations should consider upgrading them to what might be called "international historic sites." (Individual nations are not allowed to own parts of the space environment, including the planetary bodies, according to the space law treaties, as explained in Chapter 5.)

On Earth, this type of protection is afforded under the 1972 World Heritage Convention (WHC), which by 2002 had been ratified by 175 countries. Managed by the United Nations Educational, Scientific and Cultural Organisation (UNESCO), the World Heritage Fund provides financial assistance in conserving so-called World Heritage Sites. Could this somehow be extended to the planetary bodies?

The designation of parts of the planetary bodies as historic sites, heritage areas, or international parks has been suggested by several concerned authors.[49-51] In 2004, a specific proposal for a system of seven "planetary parks" on Mars was made by two European scientists, each park containing representative features of the Martian landscape. For example, a proposed Polar Park would cover part of the north pole, while Olympus Park would protect Olympus Mons "to prevent it being spoiled like Mount Everest."[52]

There have even been calls for the Moon to be designated, in its entirety, as a United Nations World Heritage Site and protected in a similar way to Antarctica, where private ownership of land and commercial exploitation are prohibited.[53] According to Rick Steiner, an academic at the University of Alaska, the Moon meets several criteria under the WHC because it contains "outstanding examples representing major stages of Earth's history," "superlative natural phenomena," and "areas of exceptional natural beauty and aesthetic importance."[53] In his opinion, "The historic value of the Moon alone—as it is the first extraterrestrial object upon which any life form from Earth has set foot—should provide sufficient justification for WHS designation."

Unfortunately, Steiner's effort to nominate the Moon, or the Sea of Tranquillity alone, was rejected by UNESCO and the U.S. Government because, he says, the WHC "does not apply off Earth." This is confirmed by UNESCO's Mario Hernandez, who points out that the Convention deals with "the protection of natural and cultural heritage situated on the territory of the countries" that have signed the Convention (Hernandez, Mario, personal communication, 10 April 2003). Indeed, the phrase "situated on its territory" appears several times in the convention,[54] and, until it is extended off world, this will remain a limitation.

Alternatively, following professional archaeological practice, it has been suggested that an internationally funded Sites and Monuments Record be compiled for the Moon and Mars, to raise awareness of the conservation issue and to facilitate future planning and commercial exploitation.[55] As part of the proposal, the sites of successful landings would include the areas around the landing points that had been explored by rovers or astronauts, while crash sites would include the point of impact and the debris field.

The notion that the spacecraft debris of the past few decades could be considered in terms of archaeology is an interesting point in itself. Archaeologist Greg Fewer quotes from an article in the journal *British Archaeology* concerning the potential evidence of sentient life on other worlds: "it's a great pity that the poor old Moon seems to boast nothing in the way of artifacts beyond the rubbish we've left there ourselves."[56] As Fewer points out, it is unfair to dismiss the artifacts as rubbish because that is what most archaeologists spend their time searching for and analyzing here on Earth. And taking the layman's point of view, many museum visitors already gaze with awe at preserved technology of the 20th century, from steam engines to Saturn Vs, as much as others gaze at Roman pottery and Greek statuary.

At some point, we need to decide whether discarded Apollo hardware, and other equipment deposited (at whatever velocity) on the surface of the Moon, should be classed as valuable historic artifacts or worthless trash. When Neil Armstrong threw the handle of his soil collector (to see how far it would go), he was discarding a tool no longer needed. Discarded tools from previous

Ages of Man are today displayed proudly in museum collections, so is that where this Space Age tool belongs? If this and other objects are classed as artifacts, then future tourists who help themselves to souvenirs could be accused of looting an archaeological site, but if it is defined as trash they should be commended for tidying up.

Considering the high prices commanded by sometimes quite insignificant items in space memorabilia auctions, it seems sensible to conclude that Armstrong's tool handle will never be considered worthless, either financially or culturally. The worth of some larger items has already been established: in 1993, for example, the rights to the Russian Lunokhod 1 rover and its Luna 17 landing vehicle were sold at a Sotheby's auction for $68,500, and because both spacecraft are now privately owned they can be legally sold to any other interested party.[57]

Aside from the interest surrounding the hardware, there is also supporting evidence that at least two of the Martian landing sites are already of some cultural interest, in that they have been named in honor of individuals: Viking 1 is known as the Thomas A. Mutch Memorial Station after the NASA associate administrator in charge of the Viking 1 imaging team, and the Mars Pathfinder site is known as the Carl Sagan Memorial Station after the planetary scientist and astronomy popularizer. In addition, in January 2004 the landing site of the Mars Exploration Rover Spirit was named the Columbia Memorial Station in honor of the astronauts who died in the Columbia accident (see Fig. 65).[58]

The immaturity of tourism as a space application tends to suggest that the potential for pollution or damage is not of immediate concern. However, the potential for irreversible damage might have to be faced relatively soon if plans to send remotely controlled rovers to the Moon reach fruition. In 1998, for example, a U.S. company (Lunacorp) announced plans to launch a privately funded spacecraft, carrying "intelligent" robot rovers to the Moon. Company publicity implied that the rovers were to be used for entertainment and scientific purposes or, as Lunacorp's President David Gump put it, "to trek among historic Apollo sites." One can only hope that some assurance that the rovers and their operators will respect the historic nature of the sites can be secured before Armstrong and Aldrin's footprints are obliterated by tyre tracks![59]

The status of this and other similar projects is uncertain, but this is one aspect of what could be called "remote tourism" that is not currently technology-limited. The only thing missing from the equation is funding. Although a remotely controlled vehicle might be preferable to crowds of lunar-booted tourists, initiatives such as this should be carefully monitored, and possibly even regulated under international space law.

Reports at the time of writing suggest that the first commercial lunar mission could be TransOrbital's TrailBlazer probe, which is designed to

Fig. 65   The landing site of the Mars Exploration Rover "Spirit"
was named the Columbia Memorial Station in honor of the
astronauts who died in the Columbia accident—a sign that our
earthbound culture extends as far as Mars.

orbit the Moon and, later, crash a capsule containing anything from gift certificates and business cards to cremated remains onto the surface.[60,61]

Another example of a potential extraterrestrial development concerned an
idea from researchers at the University of Tokyo for the installation of cemeteries on the Moon.[62] Apparently, the incentive for the plan was the increasing difficulty in finding land for cemeteries here on Earth. The authors of the
paper suggested that the idea could encourage space tourism, but that
remains to be seen.

INDUSTRIAL DEVELOPMENT

Beyond the relatively benign invasion of the space environment represented by tourism, one must also consider the potential for industrial development. The example most often written about is the mining of mineral and
other resources from the Moon, Mars, or asteroidal bodies.[63–65]

Helium-3, which is deposited on the planetary bodies by the solar wind, is seen as one of the most important potential products of extraterrestrial mining because of its use in still-to-be-developed nuclear fusion reactors, here on Earth or elsewhere. On the Moon, for example, it could be found in the first 3 m of lunar regolith across the whole lunar surface. The low concentrations of only several parts per billion mean that it would have to be collected by extensive strip-mining techniques, which would have a significant and visible effect on the lunar surface. It has been suggested that to mine a single kilogram of helium 3 would require the collection and processing of 250 tonnes of lunar regolith.[66]

The impact of such a mine from an aesthetic and cultural point of view could be considerable, because the Moon is easily visible from Earth.[67] This is not to say that strip mining of the near side should necessarily be banned; it is more a matter of determining to what extent it should be limited. For example, if mines are to be established, should their size be limited so as not to be visible with the average naked eye, thus protecting the cultural heritage of "the man in the moon"? Or should they be small enough to remain invisible through the average amateur telescope (a far more difficult requirement), or for the benefit of future lunar tourists from a low-altitude lunar orbit? These and other questions will one day be asked, not out of academic interest as they are now, but because answers are required by governments or regulatory bodies.

In an attempt to show how damaging strip mining could be to science, astronomers have cited the example of the Martian moon Phobos, which is smaller than a city the size of London. An open cast mine of a size typical on Earth could destroy its unique groove system forever (see Fig. 66).[50]

In addition to destroying large-scale physical features of interest to science, it has also been suggested[50] that the surface layers of a body could be contaminated by industrial effluents and that pollutants containing radioactive elements could render radioisotopic dating of the natural materials ineffective. Had a 1959 U.S. Air Force proposal to detonate a nuclear device on the Moon, "as a warning to the Soviet Union,"[53] been acted upon, science might already be suffering the consequences.

On the positive side, large-scale mining of helium 3 is unlikely until fusion techniques are perfected and commercialized by the terrestrial power industry. However, it is unlikely that helium 3 would be mined on its own because it would be more efficient to mine it simultaneously with oxygen, which is chemically bound in the rocks and soil of the moon.[68] Oxygen, of course, has a long-celebrated use as a rocket propellant (oxidizer), especially when combined with hydrogen as a fuel, and this might bring forward the development of the lunar mine. Indeed, it might also be possible to mine hydrogen in parallel with oxygen, which would be too useful a benefit to ignore.

Fig. 66   A consequence of industrial development in the solar system? (An aerial view of a power station and open-cast mine superimposed, to scale, on the Martian moon Phobos).

China's vice minister of COSTIND, Luan Enjie, summed it up well when he said "The Moon has become the focal point wherein future aerospace powers contend for strategic resources. The Moon contains various special resources for humanity to develop and use." He made particular mention of helium 3, which he said would "help change the energy structure of human society." In his opinion, the prospect of using lunar minerals and energy resources for the sustainable development of human society was "the most important driving force for [a] return to the Moon."[48]

There is, however, another potential problem with lunar mining: the consequent release of gases formerly bound within the regolith. If not collected and processed, these gases could pollute the pristine lunar atmosphere for kilometers around the mining site,[68] which could have damaging effects on lunar science.

Although to a first approximation and from a practical point of view the Moon has no atmosphere, it is a different matter from a scientific perspective, because very small amounts of gas are detectable. According to measurements made by the Lunar Atmospheric Composition Experiment (LACE) deployed by Apollo 17, the predominant day-side chemical species is carbon dioxide (CO_2), and the predominant night-side gas is neon. Total atmospheric masses of the main constituents are estimated as follows: 5200 kg of CO_2, 2000 kg of carbon monoxide, 640 kg of methane, and less than 500 kg of neon and argon.[41]

Naturally, to a planetary scientist, all atmospheres are unique, and the lunar atmosphere is no exception: the fact that it is "collisionless"—because of its low density—makes it, by definition, an exosphere (as defined in Chapter 2).

The Moon's atmosphere is derived from a combination of natural sources—chiefly outgassing from the surface and material deposited by the solar wind—at a rate of only 50 g/s for the entire surface. This means that any lunar-base activities will modify the natural atmosphere if gas is released at a comparable rate, a rate that might be expected by only low levels of surface activity, industrial or otherwise. Possible artificial sources include rocket exhausts, the venting of pressurized facilities, and the processing of lunar materials.

To place the lunar atmospheric mass—of no more than a few tonnes—in perspective, data from Apollo indicate that a lunar module landing and ascent added some 3000 kg of atmospheric gases, while the return of the CSM to Earth added about 5000 kg.[41] Although these additions were short lived, largely because of the clearing action of the solar wind, continual deposition of these gases in some future lunar development scenario would produce a permanent modification of the lunar atmosphere. Depending on the extent and type of development, this might result in a localized increase in atmospheric absorption and light scattering, which would have an impact on astronomical observations from the lunar surface.

Although the effect of a small lunar base might be deemed insignificant, compared with the advantages, the Apollo data show that even a single spacecraft movement has at least a transient effect. It would therefore be sensible from a scientific point of view to fully characterize the lunar atmosphere before significant industrial development takes place.

From a commercial point of view, anything mined on the Moon for use on Earth will, by virtue of distance, be more expensive than the same substance mined on Earth. However, by the time it becomes economical to mine the Moon, there might be issues of resource depletion on Earth to factor into the discussion. Indeed, the Moon might one day be the only practical source of a given material. Hopefully, by that time, methods of minimizing the environmental impact of industrial development will have become an integral part of licensing procedures and business plans, to the extent that lunar developers might find it more efficient to incorporate landscaping and landfill techniques into the industrial process than to try and tidy up later.

Again, from a commercial point of view, the mining of water ice believed to exist at the lunar poles is likely to have an impact on potential developers as well as on the lunar environment. As one commentator has pointed out, "It would be terrible if one company managed to get the right of non-interfered use of all the lunar ice."[68] So, again, some form of regulation appears to be in order.

Fig. 67   Mining Saturn's outer ring may be a product of Alfred Bester's fertile imagination, but it prompts us to set limits to industrial development.

If a business case can one day be made for this or any other type of industrial development, it would be advisable to have proved the need for an environmental impact statement in advance. The alternative would be an environmentally damaging development free-for-all, which could have both scientific and cultural/aesthetic repercussions.

The ultimate question here is whether we consider the mineral constituents of the planetary bodies simply as resources to plunder, as we do on Earth, or accord them a degree of protection. A key difficulty is deciding how much "damage" to the space environment is allowable in the name of progress.

Once again, a tale of science fiction provides an illustration of the concern. Alfred Bester's novel *The Computer Connection* describes a future in which the outer ring of Saturn has been mined "for some kind of better building aggregate"[69] (as illustrated in Fig. 67). It might well be far fetched, even tongue in cheek, but intentional exaggeration is a powerful tool in highlighting potential danger.

TERRAFORMING

Perhaps the ultimate change to a planetary environment is that offered by terraforming, the putative science and technology of altering a planet's environment to make it more Earth-like. The concept first appeared in Olaf Stapledon's book *Last and First Men*, published in 1930,[70] but was given the name "terraforming" by Jack Williamson in *Collision Orbit*, a short story that appeared in a 1942 edition of *Astounding Science Fiction*.[71] Carl Sagan pioneered the discussion of terraforming in the technical literature by publishing an article on Venus in a 1961 issue of *Science*,[72] and by 1976 NASA had published the proceedings of a workshop examining the feasibility of terraforming Mars.[73]

The danger, of course, is that future attempts to terraform Mars—the most likely candidate—could destroy existing but undiscovered life-forms, as well as changing existing landforms and other physical features (Fig. 68). To what extent this is important depends upon one's viewpoint, but it has been suggested that whereas life-forms and ecology are considered sacrosanct, the inherent beauty of geology and geomorphology is not always accorded the recognition it deserves.[1]

Although this might be true, there are those who question the point of going to Mars, if it is not to benefit mankind: "why spend all that money if we're not going to get anything out of it?," they ask. One might equally ask why we spend money on pure science—why, for instance, we study subatomic particles and the constituents of distant stars. These questions are all part of a continuing philosophical discourse on the meaning of life and the reason for our existence.

Perhaps in deciding how to treat a potential future home, we should refer to our experience on Earth and ask whether we would do it in our own back yard. Would we sanction, for example, the construction of condominiums on the mesas of Grand Canyon or the mining of the Giant's Causeway for patio materials? Probably not.

In the United Kingdom, for example, the horticultural community has realized, rather too late, that the decimation of peat bogs and limestone

Fig. 68 Clouds in an atmosphere and water in Valles Marineris are signs of terraforming Mars in this vision of the future (or is it a nightmare?) by artist David Hardy.

(© David A. Hardy/www.astroart.org)

pavements has had a detrimental impact on the natural environment. The fact that some British gardeners now have fertile soils and attractive rockeries is poor compensation for the national loss. It is logical to carry this type of experience with us as we strive to explore and develop other planetary bodies.

At the most basic level, terraforming is concerned with changing the environment of a planet, which, in its original form, is inhospitable to human life or simply uninhabitable. This implies that a planet is only considered valuable if it is hospitable to humans, but this goes against the flow in the development of environmental awareness. Some believe that now we are able to appreciate the value of Antarctica, the Sahara, and other "uncomfortable" terrestrial environments, we should be expected to extend this to the planetary bodies.[74]

ARE WE TRASHING THE SOLAR SYSTEM?

Today, as we look forward to further manned lunar missions and subsequent development, the status of the lunar surface and its orbital environment should be of greater concern, not the least with regard to the safety of future travelers. By extension, the effect of spacecraft impacts on the surfaces of other planetary bodies and the formation of planetary orbital debris should be of similar concern to the wider space community.

To most people, a few footprints, wheel tracks, and auger holes on the lunar surface are more than acceptable considering the science return. Indeed, they are seen by many as a celebration of mankind's triumph over adversity and a symbol of our conquest of the Moon. And from a cultural perspective, the first lunar boot print is considered a seminal photographic image in the history of the human race; why else would it appear on the covers of so many books and magazines?

However, in all of the words that have been written on the Apollo program, little mention has been made of the fact that the astronauts "threw out the trash" before they headed back to lunar orbit. And more than 30 years on, few realize that the ascent stages of six lunar modules were allowed to crash onto the lunar surface. Although it is still "politically incorrect" to suggest that these impacts despoiled a pristine, natural landscape and represent potential hazards for future explorers, the fact remains that future lunar exploration has the potential to create far greater impacts and hazards.

This begs the question of our collective attitude towards the space environment and the degree to which we should regulate its use and protect it for future generations. The following chapter discusses the legal aspects of protection of the space environment.

REFERENCES

[1]Williamson, M., "Planetary Spacecraft Debris—The Case for Protecting the Space Environment," *Acta Astronautica*, Vol. 47, Nos. 2–9, 2000,

pp. 719–729; also International Academy of Astronautics, Paper IAA-99-IAA.7.1.01, 1999.

[2]Taylor, J. W. R., and Allward, M., *Eagle Book of Rockets and Space Travel*, Longacre Press, London, 1961, p. 143.

[3]Reeves, R., *The Superpower Space Race*, Plenum, New York, 1994, pp. 54–58.

[4]Johnson, N., "Man-Made Debris in and from Lunar Orbit," International Academy of Astronautics, Paper IAA-99-IAA.7.1.03, Oct. 1999.

[5]Williamson, M., "The Early Development of Space Science Satellites and Planetary Probes," Inst. of Electrical Engineers, July 1996.

[6]Britt, R. R., "Earth's Newest Satellite Could Hit Moon Next Year," Space.com [cited 12 Sept. 2002].

[7]Johnson, N., "Man-Made Debris in and from Lunar Orbit," *Earth Space Review*, Vol. 9 No. 4, 2000, pp. 57–65.

[8]Williamson, M., "Satellite Moonshot: A Giant Leap for Hughes," *Space and Communications*, Vol. 14, No. 5, Sept.–Oct. 1998, pp. 22–25.

[9]Skidmore, M. R., "Lost (and Found) in Space," Analytical Graphics Insurance Symposium, Nov. 1998.

[10]Johnson, N., "The Passivation of Orbital Upper Stages, A Lesson Not Yet Learned," International Astronautical Federation, Paper IAF-97-V.5.07, Oct. 1997.

[11]Kruger, H. et al., "Detection of an Impact-Generated Dust Cloud Around Ganymede," *Nature*, Vol. 399, 10 June 1999, pp. 558–560.

[12]Reeves, R., *The Superpower Space Race*, Plenum, New York, 1994, p. 327.

[13]"NASA Mission Report—Mariners Six and Seven," MR-6 (10/29/69) *Mars: The NASA Mission Reports Volume One*, edited by R. Godwin, Apogee Books, Burlington, Canada, 2000, p. 63.

[14]Reeves, R., *The Superpower Space Race*, Plenum, New York, 1994, pp. 357–359.

[15]Williamson, M., *The Cambridge Dictionary of Space Technology*, Cambridge Univ. Press, Cambridge, England, UK, 2001, p. 136.

[16]Tennen, L. I., "Evolution of the Planetary Protection Policy: Conflict of Science and Jurisprudence," World Space Congress Committee on Space Research, COSPAR 02-A-1797, Washington, DC, Oct. 2002.

[17]Stabekis, P. D., *Report: COSPAR/IAU Workshop on Planetary Protection*, edited by J. D. Rummel, Committee on Space Research, Paris, 2002, Appendix C.

[18]Rummel, J. D., *Report: COSPAR/IAU Workshop on Planetary Protection*, Committee on Space Research, Paris, 2002.

[19]Sterns, P. M., and Tennen, L. I., "Protection of Celestial Environments Through Planetary Quarantine Requirements," *Proceedings of the 23rd Colloquium on the Law of Outer Space*, AIAA, New York, 1980, p. 111.

[20]Uhlir, P. F., and Bishop, W. P., "Wilderness and Space," *Beyond Spaceship Earth: Environmental Ethics and the Solar System*, edited by E. C. Hargrove, Sierra Club Books, San Francisco, 1986, p. 202.

[21]Clarke, A. C., "Before Eden," *The Best Science-Fiction Stories*, Hamlyn, London, 1977, pp. 500–509.

[22]Caidin, M., *War for the Moon*, E. P. Dutton, New York, 1959, p. 129.

[23]Noever, D., "Earth Microbes on the Moon," *Science*@NASA/Space Science News website, URL: http//:science.nasa.gov/default.htm [cited 1 Sept. 1998].

[24]Harland, D. M., *Exploring the Moon*, Springer-Praxis, Chichester, England, UK, 1999, p. 45.

[25]Phillips, C. R., "The Planetary Quarantine Program: Origins and Achievements," NASA SP-4902, 1974/5 (U.S. GPO stock no. 3300-00578).

[26]Rummel, J. D., "Strep, Lies (?), and 16 mm Film: Did S. Mitis Survive on the Moon? Should Humans be Allowed on Mars?," Abstract for 2004 Astrobiology Science Conference, *International Journal of Astrobiology*, 2004.

[27]Godwin, R. (ed.), *Mars: The NASA Mission Reports Volume One*, (NASA Viking Press Kit no.75-183) Apogee Books, Burlington, Canada, 2000, pp. 165–166.

[28]Sterns, P., "The Scientific/Legal Implications of Planetary Protection and Exobiology," International Academy of Astronautics, Paper IAA-99-IAA.7.1.06, Oct. 1999.

[29]Sterns, P. M., and Tennen, L. I., "Current United States' Attitude Concerning Protection of the Outer Space Environment," *Proceedings of the 27th Colloquium on the Law of Outer Space*, AIAA, New York, 1984, pp. 398–403.

[30]Sterns, P. M., and Tennen, L. I., "Recent Developments in the Planetary Protection Policy: is the Outer Space Environment at Risk?," International Inst. of Space Law, IISL-89-040, Oct. 1989.

[31]Debus et al., 1998 quoted in Sterns, P. M. and Tennen, L. I., "Recent Developments in the Planetary Protection Policy: is the Outer Space Environment at Risk?," IISL-89-040, 1989.

[32]Bohlmann, U., "Planetary Protection in Public International Law," IAF IAC-03-IISL.1.05, Oct. 2003.

[33]Bates, J., and Wiseman, T., "Scientists Replicate Mars Life Experiment," Space Log, *Space News*, Vol. 14, 17 Nov. 2003, p. 10.

[34]David, L., "Japan's Nozomi Mars Probe Stirs Contamination Qualms," Space.com [cited 3 July 2003].

[35]Hall, K., "Hope Lost, Japan Abandons Mars Probe," Space.com [cited 9 Dec. 2003].

[36]Hoyle, F., and Wickramasinghe, C., *Lifecloud: The Origin of Life in the Universe*, J. M. Dent and Sons, London, 1978.

[37]Hoyle, F., and Wickramasinghe, C., *Diseases from Space*, J. M. Dent and Sons, London, 1979.

[38]Lupisella, M., "Humans and Martians," *Earth Space Review*, Vol. 9, No. 1, 2000, pp. 50–60.

[39]Space Science Board, National Academy of Sciences, *Biological Contamination of Mars: Issues and Recommendations*, National Academy Press, Washington, DC, 1992, pp. 47–49.

[40]Britt, R. R., and David, L., "Mars Rovers Getting New Software to Speed Travel Time," *Space News*, Vol. 15, 1 March 2004, p. 23.

[41]Vondrak, R. R., "Environmental Modification by Lunar Base Activities," International Academy of Astronautics, Paper IAA-89-633, Oct. 1989.

[42]Fiennes, R., *Mind over Matter: the Epic Crossing of the Antarctic Continent*, Sinclair-Stevenson, London, 1993, pp. 158, 159.

[43]Williamson, M., "Seeing Red," Analysis, *IEE Review*, Vol. 50, No. 2, Feb. 2004, pp. 22–25, URL: http//:www.nasa.gov.

[44]"Apollo Spacecraft Cleaning and Housekeeping Procedures Manual," NASA, MSC-000 10, 12 May 1969, p. 3.

[45]"Outbound Lunar Biological and Organic Contamination Control: Policy and Responsibility," NASA Policy Directive 8020.8A, Washington, DC, 12 May 1969.

[46]Lupisella, M., "Ensuring the Integrity of Possible Martian Life," International Academy of Astronautics, Paper IAA-99-IAA.13.1.08, Oct. 1999.

[47]McKay, C., and Davis, W., "Planetary Protection Issues in Advance of Human Exploration of Mars," *Advanced Space Research*, Vol. 9, No. 6, 1989, p. 197.

[48]David, L., "China Outlines Its Lunar Ambitions," Space.com, 4 March 2003.

[49]Almar, I., Horvath, A., and Illes, E., "Phobos—a Surface Mine or an International Park?," *The Planetary Report*, Vol. VIII, No. 5, 1988, p. 17.

[50]Almar, I., and Horvath, A., "Do We Need 'Environmental Protection' in the Solar System?," International Academy of Astronautics, Paper IAA-89-634, Oct. 1989.

[51]Williamson, M., "Planetary Spacecraft Debris—The Case for Protecting the Space Environment," International Academy of Astronautics, Paper IAA-99-IAA.7.1.01, Oct. 1999.

[52]Cockell, C., and Horneck, G., "A Planetary Park System for Mars," *Space Policy*, Vol. 20, No. 4, Nov. 2004, pp. 291–295.

[53]Steiner, R. G., "Designating Earth's Moon as a United Nations World Heritage Site—Permanently Protected from Commercial or Military Uses," International Academy of Astronautics, Paper IAA-02-IAA.8.1.04, Oct. 2002.

[54]"Convention Concerning the Protection of the World Cultural and Natural Heritage," UNESCO, Paris, France, adopted 16 Nov. 1972, URL: http://whc. unesco.org/nwhc/pages/doc/dc_f15.htm.

[55]Fewer, G., "Towards an LSMR & MSMR (Lunar & Martian Sites & Monuments Records) Recording Planetary Spacecraft Landing Sites as Archaeological Monuments of the Future," Theoretical Archaeology Group, Bournemouth Univ., England, UK, Dec. 1997.

[56]Thomas, C., "Diggers at the Final Frontier," *British Archaeology*, Vol. 11, 1996, p. 14.

[57]Spennemann, D. H. R., "The Ethics of Treading on Neil Armstrong's Footprints," *Space Policy*, Vol. 20, No. 4, Nov. 2004, pp. 279–290.

[58]"Space Shuttle Columbia Crew Memorialized on Mars," NASA Press Release 04-009, 6 Jan. 2004.

[59]Williamson M., "Return to the Moon," *SPACE and Communications*, Vol. 14, No. 3, May/June 1998, pp. 23–27.

[60]David, L., "Reclaiming the Moon: Plans for a 21st Century Return," Space.com, 4 June 2002.

[61]Britt, R. R., "Fly My Stuff to the Moon: Private Mission Slated for Fall Launch," Space.com, 29 Jan. 2004.

[62]*Space News*, Vol. 8, 3–9 March, 1997, p. 2.

[63]Blair, B. R., "The Commercial Development of Lunar Mineral Resources," *Earth Space Review*, Vol. 10, No. 1, 2001, pp. 76–84.

[64]Shrunk, D. G., "The Planet Moon: Stepping Stone to Space," *Earth Space Review*, Vol. 10, No. 1, 2001, pp. 70–75.

[65]Lewis, J. S., *Mining the Sky: Untold Riches from the Asteroids, Comets, and Planets*, Addison Wesley Longman, Reading, MA, 1996.

[66]Wade, D., "Directions in Space: a Possible Future for the Space Industry," *International Space Review*, Dec. 2004, pp. 6–8 (Part I); Jan. 2005, pp. 8–9 (Part II), URL: www.satellite-evolution.com.

[67]Williamson, M., "Protection of the Space Environment Under the Outer Space Treaty," International Inst. of Space Law, Paper IISL-97-IISL.4.02, Oct. 1997.

[68]Cramer, K., "The Lunar Users Union—An Organization to Grant Land Use Rights on the Moon in Accordance with the Outer Space Treaty," International Inst. of Space Law, Paper IISL-97-IISL.4.13, Oct. 1997.

[69]Christensen, B., "Planetary Parks Proposed for Mars Conservation," Space.com, 30 Nov. 2004.

[70]Stapledon, O., *Last and First Men*, Methuen, London, 1930.

[71]Williamson, J., "Collision Orbit," *Astounding Science Fiction*, edited by W. Stewart, ASF XXIX(5), 1942, pp. 80–117.

[72]Sagan, C., "The Planet Venus," *Science*, Vol. 133, 1961, pp. 849–858.

[73]Averner, M. M., and MacElroy, R. D., "On the Habitability of Mars: An Approach to Planetary Ecosynthesis," NASA SP-414, 1976.

[74]Rolston, H., "The Preservation of Natural Value in the Solar System," *Beyond Spaceship Earth: Environmental Ethics and the Solar System*, edited by E. C. Hargrove, Sierra Club Books, San Francisco, 1986, p. 171.

Chapter 5

LEGAL DIMENSION

It was suggested in the preceding chapter that, if space exploration and associated developments are to continue successfully for the foreseeable future, some form of regulation will be necessary to protect the space environment and the resources it provides.

Here on Earth, the development of resources has often led to the degradation of the natural environment, which in many cases has encouraged the formulation of legislation to guard against it. On a global level, there have been a number of international meetings from which protocols have been derived in an attempt to safeguard the environment, and, as a global community, we now recognize their importance.

It was realized early in the Space Age that entities concerned with space-based operations, be they government or private operators, would require a degree of legal guidance. This led to a number of treaties, conventions, and agreements, which, among other things, were designed to ensure that those who wished to explore and use the space environment respected the right of others to do the same.

Although the United Nation's Committee on the Peaceful Uses of Outer Space (COPUOS) was a useful first step in providing legal guidance, it was understood that the recommendations of COPUOS and its subcommittees did not carry the force of law. This led to the creation of the Outer Space Treaty of 1967.[1]

The Outer Space Treaty (OST) provided the basic framework for international space law and the treaties that followed it. The Rescue Agreement, the Liability Convention, the Registration Convention, and the Moon Agreement are, to a large extent, elaborations of that basic treaty. The formal titles of these "space laws" are shown in the Box overleaf, along with a brief description of their coverage (the full texts of the OST and Moon Agreement can be found in Appendix B).

Does the existing body of space law provide a degree of protection for the space environment? The best way to answer this question is to analyze the relevant treaties, beginning with the OST.

SPACE LAWS

The treaties, conventions, and agreements that constitute the body of space law in its current form were negotiated and drafted under the auspices of the United Nations Committee on the Peaceful Uses of Outer Space (UNCOPUOS) and adopted by the United Nations General Assembly. However, because some space-faring nations are not signatories to all treaties, there is no fully international agreement to abide by this body of law.

The treaties are listed here, under their abbreviated titles, in the order they came into force.

OUTER SPACE TREATY (OST) 1967

The "Treaty on Principles Governing the Activities of States in the Exploration and Use of Outer Space, including the Moon and other Celestial Bodies" was adopted on 19 December 1966, opened for signature on 27 January 1967, and entered into force on 10 October 1967. As of 1 January 2005, it had been ratified by 98 nations and signed by 27.

The OST is considered the basic treaty of space law, because the subsequent four space treaties—the Rescue Agreement, Liability Convention, Registration Convention, and Moon Agreement—derive from or elaborate on its provisions.

The OST states that outer space is not subject to national appropriation by claim of sovereignty or other means, that it can be used by all states or nations for peaceful purposes, and that its use should be for the benefit and in the interests of all countries. The treaty also requires that a nation which is party to the treaty is responsible for national space activities, whether carried out by governmental agencies or nongovernmental entities, and is internationally liable for damage caused by its space objects. It also says that states and entities they supervise should avoid harmful contamination of space and celestial bodies.

RESCUE AGREEMENT (ARRA) 1968

The "Agreement on the Rescue of Astronauts, the Return of Astronauts and the Return of Objects Launched into Outer Space" was adopted on 19 December 1967, opened for signature on 22 April 1968, and entered into force on 3 December 1968. It was the first specific agreement derived from the Outer Space Treaty of 1967. As of 1 January 2005, it had been ratified by 88 nations and signed by 25 (one other "accepted rights and obligations").

The Rescue Agreement elaborates on elements of Articles V and VIII of the OST by stipulating that nations should, upon request, provide assistance in recovering space objects and/or astronauts that return to Earth outside the territory of the launching state.

LIABILITY CONVENTION (LIAB) 1972

The "Convention on International Liability for Damage Caused by Space Objects" was adopted on 29 November 1971, opened for signature on 29 March 1972, and entered into force on 1 September 1972. As of 1 January 2005, it had been ratified by 82 nations and signed by 25 (two others "accepted rights and obligations").

The Liability Convention elaborates on Article VII of the OST by stipulating that a launching state be absolutely liable to pay compensation for damage caused by its space objects on the surface of the Earth or to aircraft and for damage in space. It also provides procedures for the settlement of claims.

REGISTRATION CONVENTION (REG) 1976

The "Convention on Registration of Objects Launched into Outer Space" was adopted on 12 November 1974, opened for signature on 14 January 1975, and entered into force on 15 September 1976. As of 1 January 2005, it had been ratified by 45 nations and signed by four (two others "accepted rights and obligations").

The Registration Convention stipulates that nations should maintain a national registry of objects launched into Earth orbit or beyond and transmit certain basic information from that registry to a registry maintained by the United Nations.

Moon Agreement (MOON) 1984

The "Agreement Governing the Activities of States on the Moon and Other Celestial Bodies" was adopted on 5 December 1979, opened for signature on 18 December 1979, and entered into force on 11 July 1984. As of 1 January 2005, it had been ratified by 11 nations (Australia, Austria, Belgium, Chile, Kazakhstan, Mexico, Morocco, The Netherlands, Pakistan, The Philippines, and Uruguay) and signed but not ratified by five (France, Guatemala, India, Peru, and Romania).

The Moon Agreement elaborates on Article IX of the OST by stipulating that nations should take steps to prevent the disruption or contamination of the existing planetary environment. It also calls for the creation of a regime to oversee the exploitation of the Moon when that becomes feasible and for sharing the benefits of lunar resources as part of the "common heritage of mankind."

OTHER AGREEMENTS

The *Nuclear Test Ban Treaty* (NTB) of 10 October 1963 ("Treaty Banning Nuclear Weapon Tests in the Atmosphere, in Outer Space and Under Water") is also considered relevant to the pollution/protection of the space environment.

The *International Telecommunication Union Constitution and Convention* (ITU) of 22 December 1992 also has relevance to the use of geostationary orbit.

OUTER SPACE TREATY AND PROTECTION OF THE SPACE ENVIRONMENT

It is evident from the official title of the OST, "Treaty on Principles Governing the Activities of States in the Exploration and Use of Outer Space, including the Moon and other Celestial Bodies," that it is intended to cover much more than is currently of scientific or commercial interest. The use of the term "other celestial bodies," which includes the other planets of the solar system and their moons as well as comets and asteroids, indicates the intended all-inclusive nature of the treaty (Fig. 69).

The preamble to the treaty includes a number of phrases that echo previous UN general resolutions and are often quoted, or paraphrased, in discussions of space law:

> ... *Recognizing* the common interest of all mankind in the progress of the exploration and use of outer space for peaceful purposes,
> *Believing* that the exploration and use of outer space should be carried on for the benefit of all peoples irrespective of the degree of their economic or scientific development.

The key phrases are "common interest of all mankind," "peaceful purposes," and "for the benefit of all peoples." In an attempt to be egalitarian, then, one could say that the OST encapsulates a desire to protect the space environment because environmental degradation, in space or anywhere else, is not in the common interest or of benefit to all peoples. But is this idealistic viewpoint born out by the text of the treaty itself?

ARTICLE I

Article I is essentially a statement of equality and freedom, which intends to place all nations on an equal footing as far as the exploration and

Fig. 69 The Outer Space Treaty was designed to cover "the Moon and other Celestial Bodies," including a) Saturn and b) its moon Titan.

development of space is concerned. Its first paragraph echoes the preamble, stating that

> "The exploration and use of outer space, including the Moon and other celestial bodies, shall be carried out for the benefit and in the interests of all countries, irrespective of their degree of economic or scientific development, and shall be the province of all mankind."

Once again we see the two phrases—"for the benefit . . . of all countries" and "the province of all mankind"—that have become clichés in space law, to the extent that the combination "for the benefit . . . of all mankind" has become an informal measure for all policies and developments. Loosely speaking, if a development is broadly seen as benefiting all mankind, it is deemed acceptable.

However, in view of the developments that have already taken place in space, one has to ask whether littering planetary surfaces with discarded rocket stages and disused spaceprobes can be defined as "for the benefit of all mankind." Certainly, the intentions behind the acts are beneficial, but are the acts themselves?

Article I goes on to state:

> Outer space, including the Moon and other celestial bodies [we'll take this as read from now on], shall be free for exploration and use by all States without discrimination of any kind, on a basis of equality and in accordance with international law, and there shall be free access to all areas of celestial bodies.

Again, the treaty attempts to be egalitarian and nondiscriminatory, but is this enough to provide any real protection? Unfortunately, the deposition of man-made debris on various bodies (mainly the Moon) has made access to certain areas difficult, if not dangerous. Although no one has personally surveyed the sites of the crash landings, one can safely assume that there will be shards of metal and other materials surrounding the impact craters. A future users' guide to the Outer Space Treaty might therefore include the warning that this free access is granted at your own risk! (Fig. 70)

The mention of international law might prove to lawyers that space is not lawless, but, as far as the space environment itself is concerned, Article I offers no protection; it is entirely people oriented. The problem with the Article rests with its (admittedly laudable) democratic nature, which demands the "freedom of scientific investigation in outer space." Perhaps, in any future revision, a proviso should be attached, stating that this freedom is granted only if it does not, thereafter, limit the freedom of later investigators. In other words, the space environment should be protected

[NASA]

Fig. 70   Given the spacecraft debris on the lunar surface, should access be granted "at your own risk"? (Apollo 17 astronaut-geologist Harrison Schmitt investigates a large boulder in the Taurus Littrow valley).

for the use of future generations ... in the "common interest of all mankind" and "for the benefit of all peoples."

ARTICLE II

Article II states that outer space is "not subject to national appropriation by claim of sovereignty, by means of use or occupation, or by any other means." This is helpful in terms of protection because if "outer space" (and all that it includes) cannot be owned, it is less likely to be developed and despoiled. We shall return to the matter of sovereignty later.

ARTICLE III

Article III returns to the matter of international law, stating the following:

> States Parties to the Treaty [This is legal jargon, meaning the states or nations that are party to the treaty (i.e., those that have ratified, or given formal approval to, the treaty)] shall carry on activities in the exploration and use of outer space ... in accordance with international law, including the Charter of the United Nations, in the interest of maintaining international peace and security and promoting international cooperation and understanding.

This has no direct relevance to protection of the space environment, apart from the reference to international law and the UN Charter, which would presumably preclude the most devastating actions (such as the detonation of nuclear weapons or other weapons of mass destruction).

ARTICLE IV

In fact, Article IV covers such weapons specifically:

> "States Parties to the Treaty undertake not to place in orbit around the Earth any objects carrying nuclear weapons or any other kinds of weapons of mass destruction, install such weapons on celestial bodies, or station such weapons in any other manner."

It goes on to state that the Moon and other bodies should be used "exclusively for peaceful purposes" (another key tenet of the treaty) and forbids military bases, weapons testing, and military maneuvers (although military personnel can conduct peaceful, scientific research).

This is all good for the space environment, but the final paragraph shows a weakness with the Article as it stands: "The use of any equipment or facility necessary for peaceful exploration of the Moon and other celestial bodies shall also not be prohibited." The problem is the potentially broad interpretation of the phrase "peaceful exploration." It was certainly not considered at all "war like," during the Apollo program, to cause Saturn V third-stage boosters to impact the Moon as part of a seismology experiment; it was simply an element of the peaceful investigation of the internal structure of our nearest celestial body. This is another example of the acts being somewhat less beneficial than the intentions.

It is a matter of fact that many aspects of life considered acceptable and harmless during the 1960s, when the OST was written, have since become unacceptable and harmful. Our attitude to the Earth's environment is the obvious example. Undoubtedly, impacting large rocket stages on the Moon was acceptable in the 1960s, but surely it would be deemed unacceptable today.

This means that Article IV's acceptance of "any equipment or facility necessary for peaceful exploration" is too broad a definition. Not only does the Article allow a broad interpretation of the word "peaceful," but it also begs the definition of "necessary." Do we allow, for example, the operation of mining equipment, which permanently disfigures the lunar surface, and, if so, to what extent?

ARTICLE V

Like Article III, Article V has no relevance to the space environment. It concerns the safety of astronauts and the stipulation that nations should offer each other assistance in returning stranded astronauts to "the State of registry of their space vehicle."

ARTICLE VI

Article VI, however, returns to the matter of which space activities are allowed and who is responsible for sanctioning those activities:

> "States Parties to the Treaty shall bear international responsibility for national activities in outer space ... whether such activities are carried on by governmental agencies or by non-governmental entities, and for assuring that national activities are carried out in conformity with the ... present Treaty."

This is good in theory, but only to the same extent that the International Telecommunication Union (ITU) is *responsible* for the administration of radio frequencies; it has no powers, apart from the power of persuasion, to stop rogue users causing harmful interference, for example.

If particular states have had *responsibility* for their national activities since 1967, when the OST was enacted, does this mean that the relevant states are responsible for the space debris deposited in Earth orbit, on the Moon, and on other planetary bodies? Initially, the answer appears to be yes, because they have, by ratifying the treaty, agreed to be responsible. Are they therefore responsible for clearing it up? This is where the waters become muddied with semantics: they are morally and legally responsible, but they are liable only under certain circumstances (for example, if the state which has suffered damage deposits a claim for compensation) (Reijnen, B., personal communication, 2 May 2004). To date, no such claim has been made because states have adopted the practice of "cross-waiver of liability," meaning that each state that suffers damage by another state pays for the damage itself. Although developments in debris mitigation imply that the problem is recognized, this is a long way from any admission of liability.

Moreover, as suggested earlier, the problem of debris beyond the confines of Earth orbit is not even recognized by the majority of space practitioners. This makes it difficult to reach agreement that this type of debris is "undesirable" and, by extension, that someone has a responsibility to clear it up (or at least not produce any more). As with all environmental legislation, a sufficiently common understanding of undesirability must be arrived at before a given type of pollution can be outlawed. Obvious terrestrial analogies are the undesirability of lead in vehicle exhaust emissions and smoking in public places, both of which have taken many years to be recognized as deserving of legislation.

To give Article VI any real meaning, the implications of "responsibility" need to be discussed, agreed, and spelled out in an appendix to the treaty, or some other legal document.

ARTICLE VII

The matter of liability is addressed to some extent in Article VII:

"Each State Party ... that launches or procures the launching of an object into outer space ... and ... from whose territory or facility an object is launched, is internationally liable for damage to another State Party ... or to its natural or juridical persons by such object or its component parts on the Earth, in air space or in outer space, including the Moon and other celestial bodies."

Although this covers liability for damage to other states and their citizens, no responsibility for damage to the natural environment is included. This is probably because it is as difficult to legislate for damage to the space environment as it is for damage to an undeveloped part of the terrestrial environment. For example, is the agency responsible for the impact of a spacecraft on a planetary body (intentionally or otherwise) any more liable for the resultant environmental damage than a pilot who crashes his aircraft in a barren desert? (Fig. 71)

Quite rightly, the sanctity of human life outweighs considerations of environmental damage, but only when the two are considered head to head. In cases where human life is not at issue, it is possible to consider the environment, as terrestrial environmentalism has shown. Unfortunately, Article VII does not take this additional step.

Indeed, nor does the Liability Convention of 1972, which is a derivative of the OST. It states that the term "damage" means "loss of life, personal injury or other impairment of health; or loss of or damage to property of states or of persons ... or property of international intergovernmental organizations." There is nothing here to protect the space environment.

ARTICLE VIII

The same goes for Article VIII of the OST, which concerns the retention of ownership and control of a space vehicle, and any astronauts aboard, despite the fact that they have traveled beyond the boundaries of the launching state.

ARTICLE IX

With Article IX, however, we are at the nub of the environmental issue. Among other things, it decrees that

"States Parties to the Treaty shall pursue studies of outer space, including the Moon and other celestial bodies, and conduct exploration of them so as to avoid their harmful contamination ...."

a)

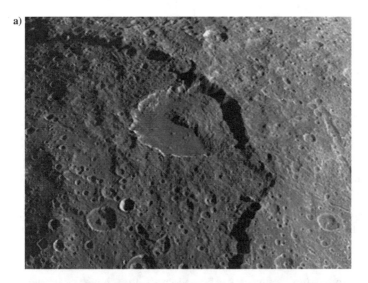

b)

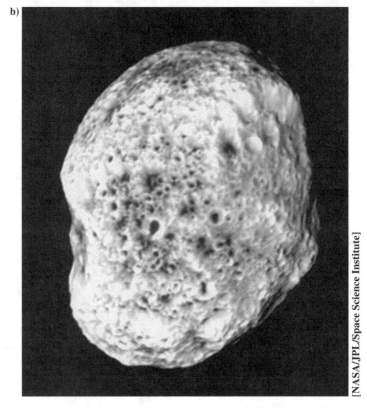

[NASA/JPL/Space Science Institute]

Fig. 71 Is it possible to damage a "barren environment," such as one of Saturn's moons? a) a landslide in the Cassini Regio region of Iapetus and b) the pumice-like Hyperion.

Theoretically, this Article should provide the space environment with the protection it deserves. Unfortunately, the phrase "harmful contamination" is open to interpretation and argument: it is likely, in the context of the treaty, that it means harmful to humans rather than harmful to the environment, especially because the Article mentions causing "harmful interference with activities in the peaceful exploration and use of outer space." The Article is more concerned with protecting activities than protecting the environment.

Apart from that, it is not specific as to what types of contamination are considered harmful. Most space professionals are happy with the concept that a planetary lander should not deliver chemical or biological contamination from Earth, because that would invalidate the science data collected by the spacecraft or, worse still, damage indigenous life-forms. However, there is less of a consensus that a spacecraft which crashes, and spreads debris over a wide area, will contaminate an otherwise pristine planetary environment, thereby potentially affecting future missions. Admittedly, there are individual examples of spacecraft that have been deorbited to protect an unexplored planetary environment—the deorbiting of Galileo in light of future missions to explore Europa being the prime example—but these are exceptions that prove the rule, not the results of a formal policy. Does this indicate a blind spot in the vision of mission planners, or is it simply an inability to see beyond the current mission—a sort of planning myopia?

ARTICLES X–XVII

The remaining Articles of the Outer Space Treaty are mainly concerned with international cooperation, the peaceful uses of outer space, and procedural matters.

MOON AGREEMENT AND PROTECTION

Although the treaties derived from the OST—the Rescue Agreement, Liability Convention, and Registration Convention—are Earth oriented in scope and add little to the cause of protection of the space environment, the most recent space treaty, The Moon Agreement of 1979, provides a step in the right direction.

It was concern for the fragility of the lunar environment, manifested in its inability to repair itself, that led to the realization that it should be protected and the production of the Moon Agreement (Fig. 72). In fact, it is clear from its formal title, "Agreement Governing the Activities of States on the Moon and Other Celestial Bodies," that the agreement was intended to apply to more than the Earth's natural satellite.

Indeed, it is clear from the treaty's preamble that, compared with the OST, another decade or so of ideas and experience of space exploration has been

Fig. 72   The subject of the Moon Agreement, shown here in an image from Apollo 16; except for the part of the image at lower left, it shows the heavily cratered far side, invisible from Earth.

incorporated. It recognizes that the Moon has an important role to play in the exploration of outer space, expresses a desire to prevent the Moon from becoming an area of international conflict, and notes the benefits that can be derived from the exploitation of the natural resources of the Moon and other celestial bodies.

Importantly, in terms of practicality, it also takes into account the need to define and develop the provisions of previous treaties in relation to the Moon and other celestial bodies with regard to further progress in the exploration and use of outer space.

So far, so good.

Article 1 sets out a few specifications regarding the applicability of the Agreement, stating that "The provisions of this Agreement relating to the Moon shall also apply to other celestial bodies within the solar system, other than the Earth," and "reference to the Moon shall include orbits around or other trajectories to or around it." There follow several statements that reiterate the intentions of the Outer Space Treaty with regard to international cooperation, peaceful uses, and the banning of weapons.

An encouraging statement, no doubt reflecting the growing environmental concerns of the 1970s, is made in *Article 4*, which says that "Due regard shall be paid to the interests of present and future generations . . . ." According to ESA lawyer Ulrike Bohlmann, this evokes for the first time in the history of space law "the principle of intergenerational equity, which is part of the more general concept of sustainability."[2] To be fair, though, the same sentence includes reference to ". . . the need to promote higher standards of living and conditions of economic and social progress and development in accordance with the Charter of the United Nations," so the cynics among us might say that it is simply the Moon Agreement's catch-all paragraph for politically correct phrases. In practice, the meaning and extent of "due regard" is likely to depend on who is considering what degree of regard is due!

However, where the Agreement gets interesting in terms of protection issues is with *Article 5*, which states in paragraph 3 that explorers should promptly make known "any phenomena they discover in outer space, including the Moon, which could endanger human life or health, as well as any indication of organic life."

Leaving aside the possibility of inorganic life, this final phrase—adapted from Article V of the OST—suggests a concern regarding an indigenous component of the space environment. Of course, considering the preoccupation with safeguarding "human life or health," it could be construed that the "organic life" was seen by those who drafted the Agreement as a threat, as opposed to something deserving protection, but we will give them the benefit of the doubt because of what follows in *Article 7*.

In what amounts to a rewrite of Article IX of the Outer Space Treaty, *Article 7.1* states the following:

> "In exploring and using the Moon, States Parties shall take measures to prevent the disruption of the existing balance of its environment, whether by introducing adverse changes in that environment, by its harmful contamination through the introduction of extra-environmental matter or otherwise."

As with Article IX of the OST, this is extremely well intentioned, but avoids the important definition of "harmful." As law professor Frank Lyall points out, "in the minimal environmental provision in the Outer Space Treaty and the Moon Agreement, we are faced with language which any lawyer would delight to interpret, depending on for whom he is appearing. The statements . . . are so vague that it is difficult to build any definite and unassailable meaning on them."[3] A possible solution, as suggested by space lawyer Patricia Sterns, is to consider any disruption "potentially harmful" and limit it to "the greatest extent possible."[4]

Of further interest is *Article 7.3*, which suggests that consideration be given to the designation of areas of "special scientific interest" as "international scientific preserves for which special protective arrangements are to be agreed upon in consultation with the competent bodies of the United Nations." This is not an end in itself because of the need for further consultation, but, suitably extended to include other celestial bodies, would make a useful addition to the Outer Space Treaty.

Indeed, it is by extension of Article 7 that some have recommended the designation of parts of the Moon as historic sites, heritage areas, or international parks (as discussed in Chapter 4). For example, it has been suggested that the lunar crater Saha, on the lunar far side, should be given special legal protection because of its proposed use in the search for extraterrestrial intelligence (SETI).[5] Being on the side of the Moon that always faces away from Earth, it is naturally shielded from radio waves emanating from the Earth, or from spacecraft in orbit around it. The crater itself is about 100 km in diameter and has a high rim, offering protection to both SETI receivers and conventional radio telescopes, and because it is located in the equatorial region it is relatively easily accessible from Earth. Since the late Jean Heidmann published his proposals for Saha in 1994, the crater Daedalus, which is almost geometrically opposite the Earth on the lunar far side, has replaced Saha as the main site of interest, and its application has been extended to encompass the whole of radio astronomy.[6]

There is some reason to be optimistic that specific areas of the far side could be protected from radio interference because, as a result of an ITU agreement of 1997, a "lunar quiet zone" was established, whereby selected frequency bands were granted protection from interference on virtually the whole far side of the Moon.[4] In effect, this mirrored ITU policy for Earth-based SETI and radio astronomy, for which dedicated frequencies were already set aside. So, although it would require specific application and significant lobbying to promote Saha, Daedalus, or any other crater as a candidate site of special scientific interest under Article 7 of the Moon Agreement, protection is a possibility.

In stating the case for protecting the lunar far side from radio-frequency interference (RFI), Claudio Maccone characterizes that part of the space environment as "a unique Natural Preserve of Humankind," adding that "we must keep it so for centuries to come by preventing any wild commercial, astronautical and military exploitation . . . in the future."[6] He is pragmatic, however, and recognizes that others would one day like to develop the L_2 Lagrangian point above the lunar far side as "a wonderful space base from which to launch [spacecraft] at little gravitational cost towards the Asteroids, Mars and the Outer Solar System Bodies" (his initial capitals).

It is also possible, as mentioned in Chapter 4, to station a communications satellite, or group of satellites, at L_2 to provide communications for

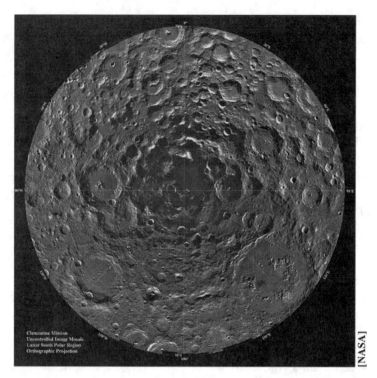

Fig. 73 If water ice exists at the Moon's south pole (imaged here by the Clementine spacecraft), who will be legally allowed to develop the resource?

settlements on the far side, which seems at odds with the concept of a lunar quiet zone. It would be up to the member of the ITU, or similar body, to revise the Radio Regulations and allocate radio frequencies to avoid interference between competing applications in much the same way as it does for Earth-orbiting satellites.

Although it seems to Maccone that the UN would be the "natural international forum" for the legal protection of the lunar far side, he concedes that this might be "just a dream!," among other things because of the potential "Moon Rush" to develop sources of water ice in the lunar south pole's Aitken Basin (Fig. 73).

Unfortunately, the needs of radio astronomers and ETI searchers might not coincide with those of other lunar interest groups, which may wish to use the Moon for other reasons. Even fellow scientists, such as lunar geologists, might not agree to bulldozing the regolith for access roads and driving piles into crater floors for observatory buildings. As on Earth, there are likely to be conflicting interests in the use of the Moon.

In a sense, *Article 8* of the Moon Agreement spoils it for protectionists by granting what are arguably excessively broad freedoms to lunar explorers and developers. Specifically, it allows them to "place their personnel, space vehicles, equipment, facilities, stations and installations anywhere on or below the surface of the Moon" and move them "freely over or below the surface...." Although the first paragraph includes the words "subject to the provisions of this Agreement," this appears to allow any future tourists—in person or via telepresence—unlimited access to the historic landing sites mentioned in the preceding chapter, unless they can be designated as protected sites in the interim.

Indeed, such designation can be viewed from two opposing directions: one in which special sites must be designated for protection in an otherwise freely developable environment; the other in which sites for exploitation and development must be designated within an internationally protected environment. It is clear that industrial developers would prefer the former, whereas scientists would prefer the latter.

Sadly, the provisions of the Moon Agreement are of academic interest only. Although the Outer Space Treaty has been ratified by 98 nations (and signed by another 27), including all of the major space-faring nations, and the Liability Convention has been ratified by 82 (and signed by another 25), the Moon Agreement has been ratified by only 11, including none of the major nations (see Box: Space Laws). As one writer dryly put it in 1990, "There are no indications that a landslide accession to this instrument is in the offing,"[7] an observation that remains true to this day.

One of the main reasons for this was the inclusion (in *Article 11*) of the phrase "The Moon and its natural resources are the common heritage of mankind...,"[8,9] which seemed to outlaw the appropriation of those resources. Since then, the phrase "common heritage of mankind" has become ubiquitous in the annals of space law and is mentioned so often that it has acquired its own abbreviation (CHM). Its ubiquity in space law papers is particularly notable because the phrase appears *only* in Article 11 of the Moon Agreement. Although the preambles to the OST, Liability Convention, and Registration Convention include the phrase "*recognizing* the common interest of all mankind in the progress of the exploration and use of outer space for peaceful purposes," CHM itself appears nowhere else in the body of space law.

However, this phrase alone is insufficient to explain the unpopularity of the Moon Agreement. A more fundamental problem was that Article 11 also returned to the matter of sovereignty, which was introduced in Article II of the OST, stating that "The Moon is not subject to national appropriation by any claim of sovereignty, by means of use or occupation, or by any other means." As if to make the point absolutely clear, it added that "Neither the surface nor the subsurface of the Moon, nor any part thereof or natural resources in place, shall become [the] property of any State, [organization, or person]."

Indeed, it even states the following:

"The placement of personnel, space vehicles, equipment, facilities, stations and installations on or below the surface of the Moon . . . shall not create a right of ownership over the surface or the subsurface of the Moon or any areas thereof."

Whereas the signatories to the OST accepted its rejection of sovereignty, and possibly delayed consideration of the implications for future space development, the direct and specific language of the Moon Agreement seems to have been a bridge too far. For those space-faring nations with a penchant for and history of commercial development, most notably the United States, this was tantamount to driving a nail in the coffin of the Moon Agreement. With no property rights, there could be no serious development and certainly no mining of materials (in fact, Article 6 had already restricted material extraction to that required for "scientific investigation").

The message in the Moon Agreement was clear: the Moon, and by extension the other planetary bodies, are the common heritage of mankind and beyond national sovereignty. It was manna from heaven for the environmentalists and a testament from hell for the developers. As a result, it attracted only a handful of disciples.

One of the many problems is that UNCOPUOS, which developed the space law treaties, operates by consensus, rather than majority voting. What this means is that all members must agree, or, put another way, a minority of states can *prevent* the approval of a proposal and its submission to the UN General Assembly. As a result, although the General Assembly approved the text of the Moon Agreement by consensus and opened it for signature, very few nations took the opportunity to actually sign it.[10]

According to one observer, the formulation of Article 11 was the outcome of "long and arduous negotiations and is generally considered a concession of developed States to developing ones."[7] If this is true, it is difficult to believe that the concession could have been anything but short term, because it discourages the development on which developed nations are, by definition, founded. Perhaps the Moon Agreement was seen as a short-term document, to be amended when off-world development became practical.

In fact, the Agreement recognized that technology would develop to the point where a degree of reconsideration was due. Article 11.5, for example, states that signatories "undertake to establish an international regime . . . to govern the exploitation of the natural resources of the Moon as such exploitation is about to become feasible."

Meanwhile, *Article 18* recommends that the situation should be reviewed either 10 years after the Agreement enters into force or, at the request of one third of signatories, any time after five years. Although this seems reasonable, a difficulty for many was the stipulation that any review would be based on the principle that "the Moon and its natural resources are the common heritage of mankind." So we are back to square one!

SOVEREIGNTY VS COMMON HERITAGE

Although it is only in the 21st century that the sovereignty of the planetary bodies is becoming a practical issue, it has already been under discussion for at least half a century. According to lawyer Bess Reijnen,[11] the matter of sovereignty in space was first discussed in 1955, in the *Journal of the British Interplanetary Society*, where it was suggested that a UN Charter "should be established for the management of outer space as an area beyond national sovereignty."[12] Although not as a direct result, the UN did establish COPUOS on an ad hoc basis in 1958, and permanently in 1961, to pursue this type of issue, which, as we have seen, is still under discussion.

In fact, the subject of ownership of the planetary bodies has already arisen, though more in the realm of media event than legal dispute. In 1980, a Californian businessman, Dennis Hope, registered a claim to the Moon, with the intention of developing an off-world real estate business. In the late 1990s, however, a German citizen, Martin Jurgens, claimed that Frederick II the Great, King of Prussia, had bequeathed ownership of the Moon to one of his ancestors "in return for services rendered back in 1756" and that he had inherited this ownership.[13] The entrepreneurial Mr. Hope had since begun to sell plots of the lunar surface, so Jurgens suggested that the German government should initiate international action against him. Although it made a good media story, it shed little light on the problem.

A similar, post-Mars Pathfinder story had it that three gentlemen from Yemen had filed a local lawsuit against NASA for infringing their ownership of Mars, purportedly handed down to them from ancestors some three millennia earlier[13] (Fig. 74). In fact, according to one author, similar ownership claims have occurred throughout recent history, including a lunar claim lodged in Chile in 1953 and a Declaration of Lunar Ownership issued by the city of Geneva, Ohio, in 1966.[14] The seriousness of such claims is not the issue, but they are useful in coaxing the otherwise staid and logical mind to stretch the boundaries of possibility. Sometimes, studying the "what if" scenarios in advance can save a lot of time and effort when the time for serious consideration arrives.

It is Dennis Hope's claims to large parts of the solar system that have been the most controversial, perhaps because they have received the most publicity. In the United Kingdom, for example, Hope's *Moon Estates* pack— typically offering "an acre of land on the Moon for £25"—has appeared

Fig. 74   A lawsuit was filed against NASA for infringing the claimant's alleged ownership of the planet Mars following the landing of Mars Pathfinder.

for sale in leading supermarkets and mail-order catalogs for several years. One catalog, published as recently as 2005,[15] offered a personalized "framed land title certificate, a map of the Moon and your mineral rights," along with similar packs for Mars and Venus (Fig. 75). Although it shared a page with the equally dubious "personalised name-a-star" certificate and the more mundane "become a [Scottish] laird or lady" gift pack, it was notable for its mention of the Outer Space Treaty. The caption read:

> The outer space treaty of 1967 elected no country could lay claim to the Moon, Mars or Venus. But it forgot to say that no individual could do so. In 1980 Dennis Hope, an American, claimed all three and filed his claim with various governments. According to current law it is totally above board. But please don't ask us how you can visit your plot of land!

Needless to say, most lawyers would agree that because no one can lay claim to the Moon the deeds and certificates are worthless novelties and purchasers are actually buying nothing. For example, in an article entitled "Moon Estates—Buyer Beware," Bess Reijnen writes "As with Antarctica and the high seas, outer space, including the Moon and other celestial bodies, may be used but cannot be sold or bought."[16] Lawyer Les Tennen is more forceful in his opinion: "Unfortunately, a disturbing trend has emerged, whereby some view the shortest path to profit is by violating space law, especially Article II of the Outer Space Treaty."[17]

Despite this, Hope apparently believes that he owns all of the moons and planets of the solar system apart from Earth and looks set to continue his business. Lawyers could—and will—continue to argue over the legality of the claim, but the issue has already migrated beyond the walls of the academic conference room. It is out in the public domain. Indeed, according

Fig. 75 You can buy land on the Moon, but do you actually own it?

to one source, more than 2.5 million people from 180 countries have "bought property" off-Earth, in sales that reached $1 million in 2003. So far, more than 400 million lunar acres have been sold, at a rate of some 1500 acres per day.[14]

The initiative—some might call it a scam—of selling tracts of lunar real estate to the general public through retail outlets such as supermarkets raises the question of what sort of developments might be allowed in future on the lunar surface. Although few people staking their claim to their own lunar acre would have any intentions of developing that real estate—whether legally theirs or not—what happens in 50 or 100 years when rival developers wish to establish a base, or a mine, or tourist facilities on the Moon? Will they be obliged to undertake a legal search for prior ownership, and would any ensuing dispute prove difficult and expensive in court? Because there is currently no internationally agreed law of the Moon, it is difficult to see how the case might be tried. Again, the lawyers tell us that this will not be a problem, because the original claims have no basis in law.

Reijnen, for example, has considered the application of the "common heritage" statement to future space tourism. She points out that under Article 11 of the Moon Agreement, a tourist could not bring a piece of Moon rock or soil back to Earth as a souvenir of their visit, because as a state, company, or private individual "you cannot own anything on the Moon, let alone take [part of it] back to Earth."[18] The concept is already familiar to terrestrial tourists, because signs in many national parks make similar stipulations.

As with much of science fact, this discussion has a precedent in science fiction. Robert Heinlein's 1949 story, "*The Man Who Sold the Moon*," concerns the establishment of the first lunar city, its legal title, and its commercial development. Although the story might well remain fiction for the foreseeable future, it would not be the first tale of space exploration and development to become fact, and those with serious interests in the field would do well to study it.

Interestingly, the ownership arguments have already been extended to the minor bodies of the solar system. In March 2000, for example, U.S. citizen Gregory Nemitz registered a claim to Asteroid 433 Eros in the Archimedes Institute's Internet registry. Then, when the NEAR-Shoemaker spacecraft landed there in 2001, he submitted an invoice for U.S.$20 to NASA for "parking and storage fees" (Fig. 76). NASA and the U.S. Department of State formally rejected the claim, but Nemitz went to court alleging legal ownership of the asteroid pursuant to the Outer Space Treaty and various provisions of the U.S. Constitution. The court dismissed the claim, but Nemitz continued to appeal.[19]

Equally interesting is lawyer Wayne White's take on rights to ownership for "improved land," a reference to settlers of the American West who, having made sufficient efforts to develop a plot of land, could claim ownership. "In fact, NASA has directly improved the asteroid, because it functioned as a base for the NEAR spacecraft's collection of scientific data," he writes, "and the site may one day be a tourist attraction or historical site, as it is the first landing site of a terrestrial spacecraft on an asteroid." If this logic is followed, the United States would also be able to claim

Fig. 76 A U.S. citizen registered a claim to Asteroid 433 Eros in 2000, then invoiced NASA for "parking and storage fees" for its NEAR-Shoemaker spacecraft.

ownership of a number of sites on the Moon, Mars, and other bodies, while other nations would own a selection of additional sites. As already pointed out, most lawyers would disagree with this notion of ownership.

Even stranger than the NEAR case was that of the Russian astrologer, Marina Bai, who filed a lawsuit against NASA with regard to its Deep Impact cometary mission (Fig. 77), reportedly asking for 8.7 billion rubles

Fig. 77 Deep impact—and another lawsuit for NASA.

($311 million) in compensation "for moral damages." She was quoted, in April 2005, as saying "The actions of NASA infringe upon my system of spiritual and life values, in particular on the values of every element of creation, upon the unacceptability of barbarically interfering with the natural life of the universe, and the violation of the natural balance of the universe." Although Moscow's Presnensky district court dismissed the case, the Moscow City Court took up the appeal, and the case continued.[20]

Another concept in international law which is relevant to this discussion is that of the global commons: the idea that there are parts of the Earth which are not subject to state sovereignty, either at all (such as the High Seas), or in the normal way of thinking of such things (such as the air, even though it is within the airspace of a state).[3] According to Frank Lyall, these areas or elements are increasingly considered to be objects held in some sort of trust for the whole of mankind, although they are not under the sovereignty of any state. "Space," he says, "would seem to be an obvious example, set aside from national sovereignty as it is." By extension, he believes, "a duty to respect the environment of a global commons could be inferred from other international environmental law, even in the absence of a clear treaty to that effect."

Regarding any future development, Lyall suggests that "the precautionary principle" (the better-to-be-safe-than-sorry approach) should be applied as states license and supervise activities in space. "If the international community is willing so to do," he adds, "it would be desirable that questions of orbital debris and of the potential contamination of or harm to celestial bodies, including the Moon, were to be tackled in a single document."

Is there ever likely to be such a document? That is a difficult question to answer.

TERRESTRIAL ANALOGY

The analogy with the terrestrial environmental is useful. Here on Earth, we are required to ensure that our vehicles do not pollute the air and that our refuse does not contaminate the water supply. Surely it would be possible to mirror our terrestrial legislation to protect the space environment.

Bess Reijnen believes it would. Citing the 1972 UN Conference on the Human Environment and the 1979 Convention on Long-Range Transboundary Air Pollution, which were designed to protect the terrestrial environment, she believes that it ought to be possible to establish "a similar convention or declaration for the prevention of the pollution of outer space."[21]

Some 114 states and nongovernmental organizations (NGOs) participated in the 1972 Stockholm conference, which among other things produced the concept of sustainable development and led to the establishment of the UN Environment Program (UNEP). This in turn led to the 1992 United Nations

Conference on the Environment and Development (UNCED), held in Rio de Janeiro, and its Agenda 21, an action program directed at sustainable development and the proper use and management of environmental resources (see Box: Example Environmental Conferences). It included 27 principles and 300 recommendations on poverty, water, biodiversity, climate change, ocean protection, technology transfer, and so on.[22]

EXAMPLE ENVIRONMENTAL CONFERENCES

1972: United Nations Conference on the Human Environment, Stockholm (the environment became an international issue)

1979: Convention on Long-Range Transboundary Air Pollution

1988: Vienna Convention for the Protection of the Ozone Layer

1992: United Nations Conference on the Environment and Development (UNCED), Rio de Janeiro. The first UN Earth Summit attracted 50,000 delegates, including 103 heads of state. Governments adopted Agenda 21, a global action plan for sustainable development containing over 2500 recommendations for action.[30] Since Rio-1992, three UN Framework Conventions have been adopted:

1) Convention on Climate Change (adopted in 1992 and further cemented by the Kyoto Protocol; aims to stabilize greenhouse gases in the atmosphere);

2) Convention on Biological Diversity (came into force in 1993; recognized that preserving biological diversity is "a common concern of humankind and an integral part of the development process"); and

3) Convention to Combat Desertification (came into force in 1997).

1997: Second Earth Summit agreed on program of action

2002: World Summit on Sustainable Development, Johannesburg

For example, Principle 21 of the UNCED Declaration states the following:

> States have, in accordance with the Charter of the United Nations and the principles of international law, the sovereign right to exploit their own resources pursuant to their own environmental policies, and the responsibility to ensure that activities within their jurisdiction or control do not cause damage to the environment of other States or of areas beyond the limits of national jurisdiction.[3]

As Frank Lyall points out, it is the clause on responsibility that is of particular interest by analogy with protection of the space environment.[3]

According to Alain Bensoussan of the French space agency CNES, the Rio Conference reaffirmed the principle of "sustainable development," which

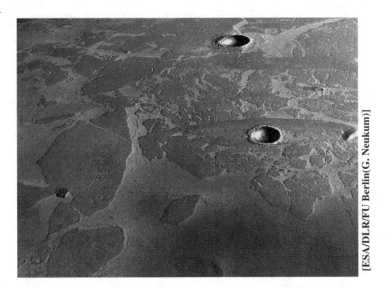

Fig. 78   Do we have a responsibility to protect this environment? This image from Mars Express appears to show a dust-covered frozen sea near the Martian equator.

was defined in the Bruntland Commission's Report of 1987 as "development that meets the needs of the present without compromising the ability of future generations to meet their own needs."[22] Obviously this is a pattern that could readily be adapted to the protection of the space environment.

More recently, the Johannesburg Earth Summit of 2002 finalized the Plan of Implementation regarding the five key headings of water, energy, health, agriculture, and biodiversity, and French President Jacques Chirac called for the creation of a world environmental organization.[22] Again, this provides a useful model for the needs of the space environment.

One of the potentially most damaging threats to the terrestrial environment is that of nuclear weapons, and it is in this regard that, according to Lyall, "environmental concern, to the point of imposing environmental duty, has become a matter of law." Thus in its Advisory Opinion of 1996 on the "Legality of the Use by a State of Nuclear Weapons in Armed Conflict" the International Court of Justice expressly states the following:

> The Court recognizes that the environment is under daily threat and that the use of nuclear weapons could constitute a catastrophe for the environment. The Court also recognizes that the environment is not an abstraction but represents the living space, the quality of life and the very health of human beings, including generations unborn. The existence of the general obligation of States to ensure that activities within their jurisdiction and control respect the environment of other States

or of areas beyond national control is now part of the corpus of international law relating to the environment.[3]

The idea of a general duty to the environment, and not just to the personnel or property of other states, has therefore been accepted, says Lyall. Although he admits that the phraseology used by the Court has an anthropocentric reference, he suggests "that it need not be so confined." Harm done to "generations unborn," he says, "could well include the degradation of our space environment, both near and far."

Once again, by analogy with terrestrial concerns, many believe that space-based weapons would be damaging to the space environment. One suggestion, made in 2002, was to convene "an emergency treaty conference [for] the Space Preservation Treaty—an international agreement [to] ban all space-based weapons."[23] The analogy was drawn with the signing, by 122 nations, of the Convention Banning Land Mines, known as the Ottawa Convention, in 1997.

Unfortunately, to date, it seems that all of the space environment has going for it, in terms of protection, is a catalog of analogies.

PROTECTION UNDER LAW?

Even a cursory analysis of international space law—specifically the Outer Space Treaty and the Moon Agreement—shows that, although in general it is well meaning, it provides insufficient protection for the space environment (Fig. 79).[24]

[NASA]

Fig. 79　When exploration turns to development, what protection will space law offer the lunar environment?

According to lawyer Bess Reijnen, who has made an in-depth study of the space treaties,[8] they do not even help with the well-recognized problem of orbital debris: she concludes that "no immediate solution follows from the obligations formulated in these treaties, neither for the clean-up . . . of existing space debris, nor for the creation of [future] space debris."[21]

It would be surprising, therefore, if they provided for the protection of the planetary bodies. According to one source, one of the problems of the OST is the fact that the Moon and other celestial bodies are "lumped together with outer space . . ."; another is "the lack of adequate enforcement capabilities."[25] It even suggests that "the treaty fails even in establishing the minimum acceptable standards of conduct," largely because of the vagueness of its terminology, particularly regarding "harmful contamination."

Unsurprisingly, there are others who criticize the OST for not specifically legislating for development. According to policy analyst Karen Cramer, "The Outer Space Treaty leaves a sovereignty vacuum. No one has sovereignty over outer space, therefore no one can grant property rights. Without property rights there is nothing to prevent anarchy among potential users of lunar territory. Anarchy is not conducive to large investments of capital."[26]

Cramer cites a proposed solution in the form of a Lunar Users Union (LUU), which would deal with lunar territory the way the ITU deals with orbital territory and spectrum resources. Recognizing that the needs of astronomers, geologists, developers, and other users are not necessarily compatible, the LUU would regulate the use of the Moon to ensure noninterference between users, with rights being granted for specified activities in specific areas. Rights would be granted on a first-come, first-served basis—the basis historically preferred by U.S. delegates to the ITU regarding resources for satellite communications—and if rights were not exercised within a given period they would be lost.

A commercially founded complaint from potential developers is that mining companies which spend large amounts of money surveying an area cannot legally restrict a competitor from entering that area to establish its own mine. Although this is a theoretical problem, because the Moon is a big place and similar resources are likely to be available elsewhere, a lack of exploration and development rights could make financing difficult because potential investors would require assurance of their clients' ability to develop.

In recognition that there should be protected areas of the Moon, the LUU proposal suggests that "historical and scientific reserves should be set aside before any other territorial assignments are made." Thus it demonstrates an acceptance of the need for balance and a recognition of the rights of scientists and conservators of historic landing sites, for example.

The ITU model for orbital and spectrum resources is a good one in that permission to use a resource is not considered a "property right": the user does not *own* the orbital position, but has a right to use it for a given period and a right to transmit and receive certain radio frequencies at that position. The fact that the right is legally and internationally recognized makes it possible to raise financing. Interestingly, this conclusion was threatened in 1976 by the Bogota Declaration, when a number of equatorial nations, including Brazil, Colombia, Congo, Ecuador, Kenya, Indonesia, Uganda, and Zaire, signed the *Declaration of the First Meeting of Equatorial Countries of 3 December 1976* wishing to claim sovereignty over the parts of geostationary orbit above their territory. Following years of discussion (and irritation in the ITU and UNCOPUOS), it was finally put to rest in 1993, when the last surviving claimant, Colombia, implicitly rejected the Declaration.[27] If nothing else, the Bogota Declaration shows that legal disputes over property rights in space, as on Earth, can be both protracted and acrimonious.

Another proposal for legalizing ownership of the space environment is contained within the Space Settlement Initiative, which has been endorsed by certain space activist groups including the Moon Society and the Artemis Society.[28] Its stated purpose is "to enable the expansion... of the human species beyond the Earth by offering a huge financial reward for privately funded settlement." Among other things, it states that while the OST prohibits national sovereignty, it "says nothing against private property," which is the argument used by Dennis Hope and others involved in selling space real estate.

Frank Lyall is succinct in his condemnation of this argument which he says is "legal nonsense [since] 'private property' presumes a legal system!" (Lyall, F., personal communication, 7 April 2004). In fact, as Bess Reijnen points out, because treaties such as the OST are agreements between states, private property is not relevant. Private property is an issue of national law, not international law, and any national law of any State is, by definition, overruled by international law (Reijnen, B., personal communication, 2 May 2004).

Nevertheless, in an attempt to encourage privately funded organizations to develop space settlements, the Space Settlement Initiative proposes the formation of legislation that would commit the United States to grant recognition of their rights of appropriation. Interestingly, there is also a nod to protection in the suggestion that "Regulations could even include protection for sites of historical or other special importance." The Initiative is little more than a Web site* at the moment, but it gives an indication of the way space enthusiasts are thinking.

As the expected development of Moon and Mars bases is pursued in the 21st century, the protection of these environments from spacecraft debris will become as important as the current formative measures regarding

---

*www.SpaceSettlement.org.

orbital debris. Unfortunately, the lack of international agreement on the inappropriately named Moon Agreement suggests that the space environment will not be accorded the protection it deserves under the existing body of space law.

This alone supports the argument that the Outer Space Treaty itself should be amended, supplemented or otherwise reinforced to offer a degree of environmental protection that is currently lacking.[24] The problem here is that treaties tend not be amended, for fear of complicating rather than clarifying matters, and although Article XV of the OST allows any state to propose amendments to the treaty, none has yet done so. In fact, several of the space law treaties, including the Moon Agreement (Article 18), recommend a review after 10 years, but no review has ever taken place (Lyall, F., personal communication, 7 April 2004).

If there is ever to be an amendment, supplement, or appendix to an existing treaty—or indeed a new treaty—it will be up to the lawyers, assuming they agree, to decide upon the form of words. However, they would be well advised to consider that space exploration and development have evolved considerably since the introduction of the space treaties and will continue to evolve. In doing so, they will be more likely to create a treaty applicable to the current century.

However, Frank Lyall is pessimistic about improving the position for protection within the body of space law, because the agenda of the relevant law-creating agencies is "fairly full," and it would take a degree of persuasion to get the question of the space environment included on them, "particularly if the matter does not seem all that important to governments." More importantly, he says "an ineffective law, one which is generally disregarded, is not only ineffective, but also is damaging to the legal system of which it forms a part."[3]

Nevertheless, he believes that the "fundamental or basic notions" of the OST, having been ratified or signed by some two-thirds of UN member states and complied with through decades of space activity, form "part of Customary International Law, and therefore are binding even on those states which have neither signed nor ratified the 1967 Treaty." So there may be hope.

In general, lawyers who have studied existing space law conclude that assigning liability for damage to the space environment is a grey area, mainly because international agreement has yet to be reached. Moreover, compliance mechanisms or procedures based on that agreement would also have to be developed.[29] Without them, it seems likely that interpretations of existing space law would be so broad and diverse that, in the face of controversy, opposing parties could only beg to differ.

In the final analysis, there is no practical use in a law, however well written, if it cannot be applied in a given situation. And unless all space-faring nations have signed up to a treaty, it cannot be universally applied.

If, as it appears, the current body of space law can offer little protection to the space environment, those who fear for the preservation of that environment, and the continuing availability of its basic resources, must find an alternative solution.

REFERENCES

[1]Tennen, L. I., "Evolution of the Planetary Protection Policy: Conflict of Science and Jurisprudence," World Space Congress, Committee on Space Research, COSPAR 02-A-1797, Washington, DC, Oct. 2002.

[2]Bohlmann, U., "Planetary Protection in Public International Law," International Astronautical Federation, Paper IAC-03-IISL.1.05, Oct. 2003.

[3]Lyall, F., "Protection of the Space Environment and Law," International Academy of Astronautics, Paper IAA-99-IAA.7.1.05, 1999.

[4]Sterns, P., "The Scientific/Legal Implications of Planetary Protection and Exobiology," International Academy of Astronautics, Paper IAA-99-IAA.7.1.06, Oct. 1999.

[5]Heidmann, J., "Saha Crater: a Candidate for a SETI Lunar Base," *Acta Astronautica*, Vol. 32, 1994, pp. 471,472.

[6]Maccone, C., "Lunar Farside Radio Lab: a 'Cosmic Study' by the International Academy of Astronautics," International Academy of Astronautics, Paper IAA.8.3.03, Oct. 2003.

[7]Terekhov, A. D., "Review Clause of Outer Space Treaties: Reflections on the Forthcoming Review of the Moon Agreement," *Proceedings of the 33rd Colloquium on the Law of Outer Space at 41st IAF Congress*, Oct. 1990.

[8]Reijnen, B., *The United Nations Space Treaties Analysed*, Editions Frontieres, Gif-sur-Yvette, France, 1992, pp. 280,307.

[9]Uhlir, P. F., and Bishop, W. P., "Wilderness and Space," *Beyond Spaceship Earth: Environmental Ethics and the Solar System*, edited by E. C. Hargrove, Sierra Club Books, San Francisco, 1986, p. 200.

[10]Sterns, P., and Tennen, L., "Legal Aspects of Settlements on the Moon and Mars: International Legal Infrastructure and Environmental Considerations," *Proceedings of the 34th Colloquium on the Law of Outer Space at 42nd IAF Congress*, Oct. 1991.

[11]Reijnen, B., "The Pollution of Outer Space," *Environmental Law Review*, Vol. 4/5, Oct. 1993, pp. 117–121.

[12]Horsford, C., "The Law of Space," *Journal of the British Interplanetary Society*, Vol. 14, 1955, pp. 144–151.

[13]Von der Dunk, F. G., "The Dark Side of the Moon—the Status of the Moon: Public Concepts and Private Enterprise," *Colloquium on the Law of Outer Space, 48th IAF Congress*, Oct. 1997.

[14]Britt, R. R., "Lunar Land Grab: Celestial Real Estate Sales Soar," Space.com, [cited 2 Feb. 2004].

[15]*Studio Christmas Book 2005*, Preston, England, 2005, pp. 248,249.

[16]Reijnen, B., "Moon Estates—Buyer Beware," *Earth Space Review*, Vol. 10, No. 2, 2001, pp. 33–35.

[17]Tennen, L., "Article II of the Outer Space Treaty, the Status of the Moon and Resulting Issues," International Inst. of Space Law, March 2004.

[18]Reijnen, B., "Space Travel and International Space Law," *Earth Space Review*, Vol. 9, No. 4, 2000, pp. 33–35.

[19]White, W., "Nemitz vs. US, The First Real Property Case in United States Courts," International Astronautical Federation, IAC-04-IISL.4.07, Oct. 2004.

[20]"Russian Astrologist Sues NASA," Space.com Astronotes [cited, 20 April 2005].

[21]Reijnen, B., "Pollution of Outer Space and International Law," International Inst. of Space Law, IISL-89-36, Oct. 1989.

[22]Bensoussan, A., "Space: Watching Over the Planet," *CNES Magazine*, No. 17, Nov. 2002, pp. 4,5.

[23]Rosin, C., and Webre, A., "Support Efforts to Ban Space-Based Weapons," Commentary, *Space News*, Vol. 13, 22 July 2002, p. 17.

[24]Williamson, M., "Protection of the Space Environment Under the Outer Space Treaty," International Inst. of Space Law, IISL-97-IISL.4.02, Oct. 1997.

[25]Uhlir, P. F., and Bishop, W. P., "Wilderness and Space," *Beyond Spaceship Earth: Environmental Ethics and the Solar System*, edited by E. C. Hargrove, Sierra Club Books, San Francisco, 1986, p. 197.

[26]Cramer, K., "The Lunar Users Union—An Organization to Grant Land Use Rights on the Moon in Accordance with the Outer Space Treaty," International Inst. of Space Law, IISL-97-IISL.4.13, Oct. 1997.

[27]Benko, M., and Schrogl, K.-U. (eds.), *International Space Law in the Making*, Editions Frontieres, Gif-sur-Yvette, France, 1993, pp. 171,172.

[28]Bates, J., "Space Settlement Initiative," *Space News*, Vol. 14, 19 May 2003, p. 11.

[29]Sterns, P., and Tennen, L., "Protection of Celestial Environments Through Planetary Quarantine Requirements," *Proceedings of the 23rd Colloquium on the Law of Outer Space*, AIAA, New York, 1980, p. 111.

[30]"ESA at the World Summit on Sustainable Development," ESA Press Release 55-2002, Paris, France, 5 Aug. 2002.

Chapter 6

ETHICAL DIMENSION

It appears from the preceding chapter's analysis of the space law treaties that they cannot be relied upon to protect the space environment and preserve its unique attributes for future generations. However, an alternative mechanism might be available in the guise of an ethical policy for space.

To most space professionals, the thought of devising a policy of space ethics is about as appealing as a wet afternoon in winter. Experience has shown that most of them are so focused on their particular area of expertise that they have little time for such "soft subjects," and their instantaneous reaction to an ethical policy would most probably be "as long as it doesn't stop us doing what we're doing, go ahead . . . ." This, however, would not be a professional attitude.

Other professions, of which many space professionals are, by virtue of their specializations, proud to be considered members, have already developed their own codes of ethics. We are familiar, for example, with ethical codes in medicine and biotechnology—subjects that deal directly with people—and in various branches of engineering that provide products and services for customers, clients, or consumers.

A code of ethics is effectively a code of conduct: it guides professionals in what they should and should not do in their everyday professional lives; it provides a useful reference for making sometimes difficult ethical judgements; and it provides external observers with an assurance of a professional's integrity and standard of competence.[1] The existence of an ethical code shows that the relevant service providers understand their duty to the service recipients and offers the latter a degree of protection and, if necessary, reparation.

The concept of an ethical code relating to an environment is less familiar, although an enhanced appreciation of the terrestrial environment has brought about a change. For example, it would no longer be considered ethical—at least in most people's minds—to develop an industrial process that seriously polluted the atmosphere, significantly depleted the ozone layer, or rendered large tracts of land or sea uninhabitable. The Earth's environment is now accorded a degree of protection, and, if part of it is damaged, reparation measures are not uncommon.

The Rio Earth Summit of 1992 marked an interesting development in our collective responsibility for the environment, but the difficulties involved in reaching agreement on the necessary measures have shown how politics and nationalism often stand in the way of good intentions and good practice. Nevertheless, it ought to be possible to extend this philosophy of environmental protection to space.

Indeed, given the ethical codes that have already been developed, there is every reason to expect, as mankind extends its reach further into space, that it should take with it a policy of ethics which covers this phase of exploration and development. And because both exploration and development have an impact on the space environment, there is every reason to expect that, following the lead of terrestrial environmentalism, an underlying theme of the code should be protection of the space environment.

DEFINING SPACE ETHICS

A typical English dictionary defines ethics as "the philosophical study of the moral value of human conduct, and of the rules or principles that ought to govern it," and "a code of behavior considered correct, especially that of a particular group, profession, or individual."[2]

In a sense, these two definitions describe the two opposing approaches to the subject: one an academic approach that considers ethics in a theoretical, idealized manner; the other a more practical way of addressing the needs of a community. It is the second approach that is required in developing an ethical code for space.

In its simplest form, a code of space ethics would be a guide to what we should and should not do in space. Although this definition gives the impression of a subject without boundaries, that should come as no surprise because ethical considerations color almost everything we do, at one level or another, affecting us as individuals or as part of a community. Almost by definition, an ethical code for space would have a bearing on almost everything we do, or plan to do, in space.

In terms of space exploration and development, space ethics would cover, for example, the impact of our actions in space on individuals, on property, on the Earth (which already benefits to some extent from our protection), and on the space environment itself.

But because space has proved so challenging to "conquer"—a common term in the annals of space exploration—relatively little consideration has been accorded to the space environment itself, in terms of the detrimental effects of space exploration and development, and relatively few space professionals consider the subject worthy of consideration. Although few would advocate a slash-and-burn philosophy in our exploration and development of the solar system, many would choose to turn a blind eye to the

**Fig. 80   Space is really big . . . so does it need protecting? (Cone Nebula
from the Hubble Space Telescope).**

occasional pollution or degradation event in favor of budgets, project time-
lines, or personal advancement. This is human nature, and it illustrates why
protection of the space environment requires a human dimension, effectively
summarized in the term "space ethics."

However, to most people outside the space community—including otherwise intelligent and professional individuals—space is a limitless, alien void
populated by marauding black holes, a handful of barren planets, and swarms
of life-threatening comets and meteors (Fig. 80). The space environment is
hardly in need of protection, they might say; if anything, we on Earth are
in need of protection from it!

Moreover, keen though they are to protect our own, planetary backyard,
few terrestrial environmentalists consider space as part of our environment,
despite the fact that it bears a magnetic field that protects us from space radiation and contains the Earth's predominant energy source, the Sun.

This lack of consideration, often manifested as ignorance, prompts the
question of our collective attitude towards the space environment.

VALUE QUESTION

Space exploration and development have been underway since the launch
of Sputnik 1 in October 1957, and few would doubt the importance of space
activity and its impact on society. It is recognized, at least by thoughtful
pragmatists, that space activities provide employment and opportunities
for wealth creation, and so, as mooted in Chapter 2, the concept of value
must be applicable to the space environment.

In pragmatic terms, the space environment is valuable because it has a use for commercial applications. Although this value tends to be taken for granted on a day-to-day basis, it would be quickly recognized should the benefits of space be somehow removed. For example, if geostationary orbit became unusable because of a buildup of debris, there would be a significant financial impact on satellite operators.

Of course, the space environment is also valuable from a scientific perspective, and scientists have a vested interest in maintaining its relative purity (at least for the course of their study). Planetary scientists are concerned about potential contamination of planetary bodies by visiting spacecraft, while ground-based astronomers are concerned at the potential disruption to observations, at both optical and radio wavelengths, from orbiting spacecraft.

The issue for industrialists and scientists alike is that current attitudes could prejudice future activities. The potential of the debris-clogged orbit or the contaminated canyon are simply different manifestations of the same lack of understanding and appreciation. Both eventualities call for protection of the respective resource.

Meanwhile, considering the future of space tourism, it is difficult to place a value on footprints and historic sites of exploration, but if it can be done for the Earth, then it can be done for the Moon and other planetary bodies.

Even though we are still in the early stages of recognizing the practical value of the space environment, its value is also recognized in a philosophical sense, in that space provides an outlet for mankind's inherent desire to explore and conquer new environments.[3] It can even be argued that the space environment is valuable because it represents freedom, by providing an almost unlimited expanse for mankind to explore, understand and, if he so wishes, to conquer. So, if a part of that expanse—a planetary surface, for example—became inaccessible for some reason, a part of that freedom would be lost (Fig. 81).

The trouble is that "freedom" is open to interpretation and can easily be abused, as shown by the concept of frontierism, which was illustrated most notably in the expansion of the American West. According to philosopher Alan Marshall, frontierism is "not so much a social or psychological concept as an economic philosophy."[4] Once an individual has surmounted the challenges of a new situation, or environment, and carved out an existence, he says, "they deservedly call that territory or environment their own." By extension, he adds, frontierists believe that the planets and moons of the solar system are "valueless hunks of rock until acted upon by humans to produce economic value and contribute to capital accumulation." This type of person represents the archetypal "bad guy" in the plot of environmental protection.

Although those in the space community might have a more informed view than those outside, most of them are likely to need some persuading that the space environment is worth protecting *for its own sake*, for example, because

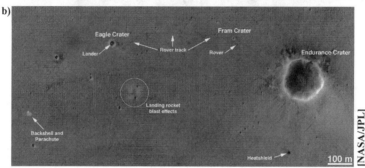

Fig. 81 Is the space environment freely available for exploration and future exploitation . . . or a wilderness that should be protected? The exploration of Mars has already left its mark: a) tracks created by NASA's Spirit rover and b) by the Opportunity rover (imaged from orbit by Mars Global Surveyor along with the rover itself, the lander and separate rocket blast effects).

parts of it might harbor simple forms of alien life, because they contain unique physical formations, or simply because they are beautiful. To the majority, aesthetics is an even softer subject than ethics!

Philosophically though, there should be no difference between an aesthetic appreciation of terrestrial objects, events, and processes and extraterrestrial ones. As one writer has pointed out, someone who appreciates rugged landscapes such as cliffs, gorges, and canyons on Earth would also appreciate their counterparts on Mars, whereas anyone who enjoys watching storm clouds move and grow would enjoy watching the storms in Jupiter's turbulent atmosphere[5] (Figs. 82 and 83).

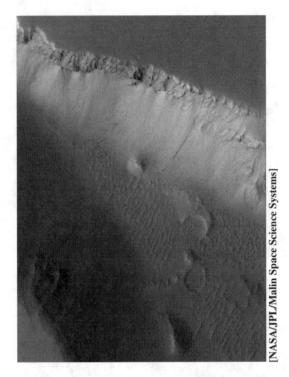

Fig. 82   Is a geological feature worthy of protection for its own sake? This Mars Global Surveyor image shows the floor and walls of a trough west of the Elysium Mons volcano. Large boulders and windblown ripples can be seen on the trough floor.

Of course, one of the fundamental problems in any value system is assigning a definitive measure of value to an item, or even a measure of relative value. Today, the Earth's immediate space environment—the orbital environment—is recognized as having a value, but the further a body is away from Earth, the smaller the value.[6]

This is mainly a matter of accessibility, or lack of it, and reflects our historical views of the world in that faraway places were typically not accorded the same respect as the homeland. These views have evolved in recent decades as a result of global tourism, and, today, even remote parts of the world are recognized as valuable (as shown by the invention of the term "ecotourism"). It is also, therefore, a matter of broadening general awareness and appreciation of the space environment.

One philosophical view of value is the following: "If a thing has formed integrity, or is worthy of a proper name, it should be respected."[5] However, as astronomer Ivan Almar points out, with regard to the space environment "whether a feature has an official name or not is almost

[NASA]

Fig. 83 Jupiter's turbulent atmosphere from Voyager 1 in March 1979, taken when the spacecraft was five million km from the planet. The Great Red Spot is visible at upper right, along with several white ovals (the smallest visible details on the original are about 95 km across).

accidental, depending on the resolution of space probe images at our disposal."[7] Does this mean, by extension, that features that have yet to be imaged have no value? The early Mars probes showed the planet as a cratered world, like the Moon or Mercury, and completely failed to detect the volcanoes and canyons of which we are now aware. Does this mean that these features were suddenly endowed with value on the day their images were received and processed? Or perhaps when they were formally named? A discussion best left to philosophers, perhaps.

A further step along the philosophical path brings one to the question of ethical rights, particularly those of indigenous extraterrestrial life-forms. The question of whether certain animate and inanimate objects have rights has exercised a number of philosophers and ethicists with interest in the subject. Martyn Fogg, for example, identifies four rival ethical theories involving different viewpoints and applies them to the morality of terraforming Mars.[8] The theories—namely, anthropocentrism, zoocentrism, ecocentrism, and preservationism—are, in essence, defined as follows.

Anthropocentrism assumes that only humans have rights because they can think rationally and act morally. Animals, plants, and microbes, for example, have no rights other than those that humans choose to give them. This is the system of ethics under which we operate now, which Fogg calls the "default system." "For the anthropocentrist," he says, "it is humanity that counts: if Mars counts more to us as a second home than as a barren desert, then ... terraforming the planet would be a moral cause."

Zoocentrism extends those moral rights beyond humans to the animal population and effectively makes our exploitation of them—for food, medicine, science, or entertainment—immoral. However, according to Fogg, it does not extend these rights to lower organisms and inanimate objects, and because higher animals are unlikely to be found on Mars the theory has nothing against terraforming.

Ecocentrism assumes not only that "all life is sacred," but also that "humans have no privileged place [within] the living world." This means that any Martian bacteria would have moral priority over humans and that terraforming Mars would, in simplistic terms, only be moral if it is a truly barren world.

Preservationism, in terms of the space environment, asserts that "the cosmos has its own values [and] its mere existence gives it not only the right to exist, but the right to be preserved" from human intervention. As such, this is the only viewpoint against terraforming. Fogg concludes that "Ideally, cosmic preservationists would like terrestrial life to stay at home, to observe and empathise with the universe, not to invade it."

This could be the case for true preservationists, but the problem with placing people in categories, like religions or political parties, is that convention does not encourage or allow them to stray outside those boxes. They become polarized in their views and blinkered against the alternatives.

Why should we not have both: a degree of protection and preservation of the space environment as well as exploration and development? Although ethical values are important, we should not get too caught up in the theory; we have to be pragmatic about the exploration and development of space.

TOWARDS AN ETHICAL CODE

Whether one's stance is pragmatic or philosophical, the logic is clear: if the space environment is valuable, it is worth protecting. The challenge is to engineer a balance between overbearing protection and unbridled exploitation. For example, how do we balance the rights of a developer to mine the Martian surface and those of a scientist to examine a pristine alien environment? (Fig. 84). Perhaps the compromise of allowing archaeologists a limited period of excavation prior to laying a building's foundations offers a suitable model.

This reference to examples, both terrestrial and extraterrestrial, highlights an important point in deriving an ethical policy. There is a danger in the discussion of ethics—perhaps because of its nature as a nonscience subject— that consideration is confined to the philosophical aspects, thus excusing those involved from providing practical solutions to the problems that

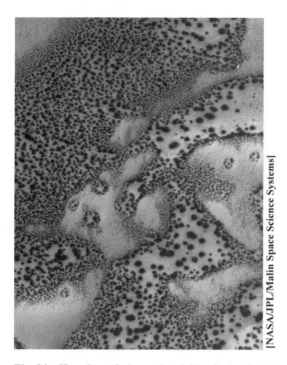

[NASA/JPL/Malin Space Science Systems]

Fig. 84   How do we balance the rights of scientists to study a pristine environment and the rights of commercial developers? The Martian south polar region from Mars Global Surveyor, showing sand dunes with the dark spots that form in carbon dioxide frost as the annual defrosting process begins.

emerge. The fact that mankind has already affected, and arguably damaged, the space environment transports the discussion beyond the philosophical realm, as illustrated by a selection of examples of mankind's impact on the space environment:

1) Project West Ford/Midas 6, 1963: clumps of 18-mm copper dipoles at 3600 km;

2) debris from spacecraft and upper-stage explosions in LEO;

3) debris from launch-vehicle separation devices in LEO and GTO;

4) microdebris in LEO (e.g., spacecraft paint and thermal insulation and metallic particles from solid-propellant motors);

5) growing population of defunct satellites in GEO-graveyard orbits;

6) impact debris of spacecraft on Moon (e.g., Luna, Ranger, Lunar Orbiter, Apollo, Lunar Prospector);

7) materials (including trash) ejected from Apollo lunar modules before liftoff;

8) impact debris of Saturn S-IVB rocket stages on Moon; and

9) similar debris (e.g., jettisoned covers) on surfaces of Venus and Mars.

So, in the same way that medical ethics concerns real-world issues, such as organ donation, assisted conception, and cloning, a policy of space ethics must evolve by addressing actual issues. Any attempt to derive a code of ethics from a philosophy is missing the point: the code must be an operational tool, not simply a list of postulates.

This implies that an ethical code for space should be based on consideration of practical examples, in effect "what we should and should not do in space." In the context of the current discussion, the question is "to what extent should we protect the space environment?" Should we regulate its use to protect it for future generations, or should we simply continue the laissez faire attitude of previous generations? It is questions such as these—the "should we" questions—that have motivated some space professionals to consider an ethical code for the future development of space.

A few example questions, in no particular order of importance, indicate the potential breadth of the discussion:

1) Should we allow adventure tourism in low Earth orbit, or will this eventually lead to an increase in orbital debris?

2) Should we allow tourists to visit the Moon, or will this lead to the pollution of a pristine environment and damage to historic lunar landing sites?

3) Should we allow the construction of orbiting advertisements visible from Earth, which would hamper astronomical observations?

4) Should we allow scientists to land, and sometimes crash, their spacecraft on the surfaces of planetary bodies without initiating a regulated spacecraft decontamination program?

5) Should we allow the terraforming of Mars without first ensuring that no indigenous, even dormant, life-forms exist?

To a large extent, the answers to these questions depend on how we value different aspects of the space environment, which suggests that an informed discussion of the value question should be a prerequisite for the derivation of an ethical code. We need to understand, and to some extent agree on, the relative values of parts of the space environment before we can decide what we should and should not do with, or to, those parts.

To be of practical use, an ethical code for space should include guidelines for the exploration and development of the space environment and for the protection of that environment for future generations.

Points of consideration should, among others, include the following[9]:

1) Protection of Earth orbits (including LEO, MEO, and GEO) as a commercial and scientific resource by formalizing debris-mitigation measures;

2) Protection of planetary surfaces (and those of smaller planetary bodies) to preserve their environments for future scientific study;

3) Protection of planetary surfaces to preserve possible indigenous life-forms;
4) Protection of historic exploration sites on the Moon and other planetary bodies; and
5) Protection of geological formations and other natural features of planetary surfaces for the enjoyment and edification of future generations (the "national park" philosophy).

One author, David Livingston, has gone so far as to propose a code of ethics comprising 12 specific principles,[10] the first four of which are prioritized as follows:

1) We will consider the effects of all off-Earth development on future generations that will live and work in space and on Earth.
2) Our business dealings in space and on Earth will be of the highest level of integrity, honesty, fairness, and ethics.
3) We are committed to ensuring a free-market economy off-Earth.
4) We agree to treat outer space with respect, concern, and thoughtful deliberation, regardless of the presence or absence of life-forms.

The fourth principle addresses the thesis of protection of the space environment and recognizes that parts of that environment might have a value in their own right, whether life exists there or not. Principle 6 extends this by supporting the establishment of "protected zones" on the Moon and other bodies.

Although it would be naive to expect that one person's initial suggestions could form a policy of space ethics for the whole of mankind, we have to start somewhere, and, as Livingston freely admits, any code should be expected to evolve as new issues arise. Contrary to the opinions of detractors, he believes that a successful code would actually promote the growth of space commerce. Moreover, says Livingston, "if the commercial space industry does not develop its own effective professional code of ethics, then government-imposed regulations will certainly fill the void. Should this occur, future development would probably be far more difficult and costly."

This is almost certainly the case. If an ethical code or policy is to be developed for space, it is important that the majority of the space community, or space profession, is intimately involved. If it is not, the profession risks having the job done for it by politicians and members of the general public, who for their own reasons might wish to place restrictions on space development, or ban it altogether.

Jacques Arnould, an ethicist with the French space agency CNES, is convinced that the space community should take "ownership" of a policy of space ethics. He believes that it is important for leaders of space policy and programs to undertake ethical studies and "take part in ethical reflection and consideration in the name of all humanity."[11] He adds that "ethically correct" is a contemporary and commonplace term, and one which space enterprise cannot ignore.

Although some believe ethical consideration of space activity to be nothing more than an intellectual exercise or "a way of sugaring the pill for public opinion on the space budget,"[11] Arnould does not subscribe to this notion. "Space development is not simply a useful activity to be regulated by codes of good will or common legislation for human activities," he says. "Space is also characterised by scientific and even mythical purposes. Through space, humanity discovers and, simultaneously, fulfils part of its own nature and the great questions which animate it, such as its origin and its destiny . . . ." In other words, it is too important to be left to the politicians!

The risk of not adopting a proactive stance is illustrated by the fate of the terrestrial nuclear power industry, which has been wounded—perhaps fatally—by a combination of scientific ignorance among the general public, effective campaigning by antinuclear groups, and bad publicity surrounding nuclear accidents. The nuclear profession, which includes scientists, engineers, and policy makers, has done too little, too late, to counter the rhetoric of the antinuclear lobby, with the result that "nuclear" is now widely considered a dirty word. In fact, the word is now so "politically incorrect" that the medical profession has all but ceased referring to its NMR body scanners as *nuclear* magnetic resonance devices, preferring the sanitized "magnetic resonance imager."

The ramifications of antinuclear campaigning within the space industry have also been significant. Research into nuclear-powered rockets (e.g., NERVA) was curtailed decades ago and, more recently, antinuclear activists in the United States forced a moratorium on the use of radioisotope thermoelectric generators (RTGs) in planetary spacecraft (staging protests at the launch of Cassini to Saturn). Although there are signs that NASA is willing to resist this and restart work on RTGs and other nuclear systems, the freedom to explore the solar system beyond the orbit of Mars has already been compromised and could be once again.

This example shows that an ethical policy must work both ways. A policy of space ethics must not only protect the space environment, but must also protect the rights of those who wish to explore and develop it. A balance must be struck.

This begs the question of who should be involved in the derivation of an ethical code for space.

WHO DECIDES?

As ever with formative issues, the questions are many, and the answers are few. One of the reasons for this is the relatively small constituency for the issue in question. At present, consideration of protection of the space

[NASA]

Fig. 85   Cassini, shown here releasing the Huygens lander towards Titan, could not have operated without its radioisotope power source.

environment is confined to a small body of concerned space professionals who consider it their duty to question the status quo.

An attempt to expand the constituency was made in 1999, when protection of the space environment was the subject of a Scientific-Legal Round Table, organized jointly by the International Academy of Astronautics (IAA) and the International Institute for Space Law (IISL) as part of the 50th International Astronautical Congress. The content of the discussion ranged from scientific and technical to legal and ethical, and some interesting papers were produced,[6,8,12–15] but the constituency for the subject remains small. The importance of promulgating these ideas, and of formulating policy, was addressed in a session of the 2002 World Space Congress.[16]

But who should consider these ideas and in what forum? It seems clear that an international consultative body should be formed to consider the relevant issues and raise awareness of the subject among the growing body of space professionals and practitioners.[17,18] Obviously, given the increasingly international nature of space exploration and development, it is important for the body to be as international as possible, ideally including representatives from all major space-faring nations.

Fig. 86 The International Academy of Astronautics (IAA) is a possible forum for discussion of an ethical code.

One suggestion would be to organize the body under the auspices of the UN, possibly affiliated with the United Nations Committee on the Peaceful Uses of Outer Space (UNCOPUOS), which has already taken an interest in space debris. An alternative might be the formation of a working group under the joint auspices of the IAA and the Committee on Space Research (COSPAR), whereas a third option might be a space agency/space industry forum.[17]

Rick Steiner, of the University of Alaska, suggests specifically "the establishment of a United Nations Outer Space Environment Commission (UNOSEC) to review and approve/reject/amend any proposed human activity in space—scientific, commercial, or other."[19] Its principle role, according to Steiner, would be to ensure "compliance with the letter and intent of the Outer Space Treaty," which itself would be enhanced by the addition of an "Environmental Protocol [that would] hold the protection of the environment of space and all celestial bodies as its highest goal." And by analogy with terrestrial affairs, he adds, any developments would be subject to the approval of "a rigorous Environmental Impact Statement." The pros and cons of these and other options might have to be discussed at some length before a decision is made.

Indeed, several space agencies have already engaged in the debate. ESA's initial contribution, for example, was led by a team from UNESCO's World Commission on the Ethics of Scientific Knowledge and Technology (COMEST) headed by Professor Alain Pompidou, the report of which was published in July 2000.[20]

It concluded the following:

> Outer space is part of the shared heritage of mankind and as such its exploration and exploitation must be freely accessible for the benefit of all mankind.

Fig. 87 The COMEST report of 2000 discussed the ethics of space policy.

The ethical approach to space is a moral principle for action . . . It must be founded on a new strategy of communication. As part of that strategy, it is imperative to leave room for a dream . . . while bearing in mind the reality of the future of space policy for the benefit of all mankind.

Perhaps unsurprisingly, it evokes one of space law's favorite principles—the CHM or common heritage of mankind principle—discussed in the preceding chapter. Because the inclusion of this phrase in the Moon Agreement is among those blamed for its failure to attract widespread ratification among space-faring nations, it should be used with some care. This is not to say that the concept of "shared" or "common" heritage is invalid, simply that certain concepts carry historical baggage.

On the positive side, the COMEST report recognizes the need to balance the requirement for control and the desirability of freedom. Indeed, as suggested earlier, ensuring a balance between protection of the space environment and the freedom to use its resources must be a major tenet of any ethical policy. Although, according to the report's conclusion, it represented only a preliminary phase, it provided a good foundation on which to build. The challenge now is to do so in an effective and pragmatic manner.

For example, a common-sense approach that addresses the basics of an ethical policy first is preferable to one that takes a fine-detail approach on a number of confined, special-interest issues. It would be advisable, therefore, to seek international agreement on a number of basic issues before producing a complex set of guidelines or policies that only a few nations would be willing to ratify. One needs only to look at the body of space law to realize that well-meaning work on behalf of the international space community is not always appreciated.

TIME IS OF THE ESSENCE

Discussions of ethical issues arising from space activities can be very broad and far reaching—in both space and time, but such discussions are little more than a way to pass the time if they remain philosophical and academic. For such discussions to be of any practical use, they must be targeted towards the design of an ethical code or policy.

Moreover, time is of the essence. It is clear from commercial developments that mankind is intent on making the space environment part of its domain. For example, the fact that an automated rover, controlled by Earth-bound theme-park patrons, could be trekking across the lunar surface within the decade makes consideration of any restrictions on its exploration relatively urgent. Even if the first such rover were to travel nowhere near the Apollo 11 landing site, to name but one historically important destination, what degree of protection could be offered to those historic first footprints from subsequent rovers and, eventually, actual tourists? (Fig. 88).

Despite the difficulties, it is important to pursue the design of and agreement on a code of space ethics with some degree of urgency. In practice, agreement on an ethical code for space might prove as difficult as agreement in space law, a topic that has been under serious discussion since the beginning of the Space Age. Nevertheless, an effort must be made now, before more serious and irreparable damage is done to the space environment.

The danger inherent in not developing an ethical code for space, or of not including protection of the space environment as a part of its foundation, has already been demonstrated by the former laissez-faire attitude towards the terrestrial environment. Space professionals must learn from history and adopt a proactive stance in protecting the space environment.

Perhaps the thought of a ethical code is too radical; perhaps we should consider a set of guidelines or a code of best practice as an initial step. But what we should *not* do is allow potentially damaging activities to continue without due consideration of the effects: a laissez-faire policy is not acceptable.

In the final analysis, although it is useful to draw analogies with terrestrial codes of ethics and to learn from their successes and failures, space demands a somewhat different philosophy, based on detailed knowledge of the space environment. Various aspects of this philosophy have had to be adopted by engineers, program managers, and policy makers to enable mankind to

[NASA]

Fig. 88   Buzz Aldrin on the Moon—the astronaut has gone home, but his historic footprints could remain for millennia. Should we—indeed can we—protect them?

explore space in the first place. Other aspects have been adopted by the people who have actually lived and worked in the space environment. It should come as no surprise that those engaged in discussions of space ethics will also be obliged to adopt a similar philosophy.

Quite what that philosophy entails is difficult to pin down, possibly because space exploration is relatively new to mankind, but it is grounded in a belief system that appears to underpin the activities of most space protagonists. They believe that space exploration is an important and worthy activity for mankind—in fact, considering our natural proclivity for exploration, a necessary activity. They know that the Earth is but a pebble on the beach of the universe, but they understand the importance of that pebble and its immediate environment. On the other hand, they see no reason why Earth should remain the only pebble of interest and seek to provide the means to live beyond what Konstantin Tsiolkovsky called (in a letter of 1911) "the cradle of mankind."

Although mankind might be years from a return to the Moon, decades from establishing an infrastructure for space tourism, and centuries from terraforming Mars, the next half-century of space exploration and development is as difficult to predict as the first was in 1957, when Sputnik 1 opened the Space Age. Had an ethical code for space been in force in the late 1950s, much of the damage to the space environment might not have occurred in the decades that followed. Given the potential for development and exploitation of the space environment in the coming decades, there can be no advantage in further delay.

As one experienced commentator has pointed out, "not only will humans travel into space ... they will also take their humanity into space."[21] We should recognize that there are many facets to this humanity: some good, which we need to encourage, and some bad, which we need to discourage.

REFERENCES

[1]Ng, J., "Engineering Ethics," *IEE Engineering Management*, Vol. 13, No. 6, Dec. 2003/Jan. 2004, pp. 30–33.

[2]*Collins Concise English Dictionary*, 3rd ed., Harper Collins, Glasgow, Scotland, UK, 1992, p. 439.

[3]Williamson, M., "Exploration and Protection—A Delicate Balance," International Academy of Astronautics, Paper IAA-01-IAA.13.1.03, Oct. 2001.

[4]Marshall, A., "Development and Imperialism in Space," *Space Policy*, Vol. 11, No. 1, Feb. 1995, pp. 41–52.

[5]Rolston, H., "The Preservation of Natural Value in the Solar System," *Beyond Spaceship Earth: Environmental Ethics and the Solar System*, edited by E. C. Hargrove, Sierra Club Books, San Francisco, 1986, p. 167.

[6]Almar, I., "Protection of the Planetary Environment—The Point of View of an Astronomer," International Academy of Astronautics, Paper IAA-99-IAA.7.1.02, Oct. 1999.

[7]Almar, I., "Protection of the Lifeless Environment in the Solar System," Committee on Space Research, COSPAR02-A-00485, Oct. 2002.

[8]Fogg, M., "The Ethical Dimensions of Space Settlement," International Academy of Astronautics, Paper IAA-99-IAA.7.1.07, Oct. 1999.

[9]Williamson, M., "Space Ethics and Protection of the Space Environment," IAF IAC-02-IAA.8.1.03, Oct. 2002.

[10]Livingston, D. M., "A Code of Ethics for off-Earth Commerce," IAF IAC-02-IAA8.1.05, Oct. 2002.

[11]Arnould, J., "Space, Ethics and Society—A CNES Study," *Earth Space Review*, Vol. 10, No. 2, 2001, pp. 58–61.

[12]Williamson, M., "Planetary Spacecraft Debris—The Case for Protecting the Space Environment," *Acta Astronautica*, Vol. 47, Nos. 2–9, 2000, pp. 719–729; also International Academy of Astronautics, Paper IAA-99-IAA.7.1.01, Oct. 1999.

[13]Johnson, N., "Man-Made Debris in and from Lunar Orbit," International Academy of Astronautics, Paper IAA-99-IAA.7.1.03, Oct. 1999. ·

[14]Lyall, F., "Protection of the Space Environment and Law," International Academy of Astronautics, Paper IAA-99-IAA.7.1.05, Oct. 1999.

[15]Sterns, P., "The Scientific/Legal Implications of Planetary Protection and Exobiology," International Academy of Astronautics, Paper IAA-99-IAA.7.1.06, Oct. 1999.

[16]"Planetary Protection Issues in Space Exploration: Science, Policy and Legal Aspects," (PPP1/G.2/IISL.5/): Joint Session Between COSPAR, IAF, and IISL, Houston, TX, Oct. 2002.

[17]Williamson, M., "Protection of the Space Environment: the First Small Steps," Committee on Space Research, COSPAR02-A-01364/PPP1-0014-02, Oct. 2002.

[18]Lupisella, M., and Logsdon, J., "Do We Need a Cosmocentric Ethic?," International Academy of Astronautics, Paper IAA-97-IAA.9.2.09, Oct. 1997.

[19]Steiner, R. G., "Designating Earth's Moon as a United Nations World Heritage Site—Permanently Protected from Commercial or Military Uses," International Academy of Astronautics, Paper IAA-02-IAA.8.1.04, Oct. 2002.

[20]Pompidou, A., "The Ethics of Space Policy," Commission on the Ethics of Scientific Knowledge and Technology, United Nations, Educational, Scientific and Cultural Organization, June 2000.

[21]Mendell, W., "Ethics and the Space Explorer," IAF IAC-02-IAA.8.1.01, Oct. 2002.

Chapter 7

SCENARIO A: DEVELOP AND BE DAMNED

Chapters 7, 8, and 9 are written from the perspective of the year 2057, the 100th anniversary of the launch of Sputnik 1 and the centenary of the Space Age. They offer three different views of how the future might turn out for space exploration and development under certain premises.

This first scenario assumes that any activities and developments are allowed without control, without an improvement in debris mitigation strategies, and without a policy for protection of the space environment.

"Sputnik's 100th Anniversary"
The Global Review
October 2057

The Global Review's space correspondent visited the Sputnik 1 Memorial Museum at the Baikonur Cosmodrome, Kazakhstan, to mark the 100th anniversary of Sputnik 1's launch on 4 October 1957.

The Space Age began 100 years ago with Sputnik 1's triumphant opening of a new frontier. Since then, there have been countless developments in space technology, but have they really met the promise of space exploration? Many experts would, emphatically, say "no."

It was in the early years of the current century that the real problems began. To most people working in the space industry today, that era is ancient history, their grandparent's era, when it took eight hours to cross the Atlantic and personal computers needed desks to support their weight. But history has much to teach us, and if we had listened to our grandparents, we might not have some of the problems we have today.

GRAVEYARD SCARE

Let us start by looking at what is still the most successful space application of all time: telecommunications satellites. We have been launching them to geostationary orbit, or GEO, almost since the dawn of the Space Age and, as everyone knows, have become totally reliant on them. Despite numerous attempts to use constellations of satellites in other orbits, GEO remains prime real estate for communications.

The problem is that geostationary orbit has a limited capacity in terms of the different radio frequencies the satellites can use, and as more efficient ways to use those limited frequencies have been developed, the satellite industry has had to update the hardware in orbit by launching new satellites. The development of broad-bandwidth optical links, using laser beams, has also persuaded operators to invest in new satellites, and because all satellites have a limited lifetime—based on equipment failures and the availability of maneuvering propellant—they have had to be replaced every 15–20 years or so. The question for industry has always been, where do we put the old satellites?

Since the late 20th century, the answer has been the graveyard orbit, which in reality is a number of orbits of slightly differing altitude above GEO. But following 80 odd years of this practice, which was recognized as a short-term solution way back in the 1980s, the graveyard is full, and no one seems clear on what to do about it (Fig. 89).

Of course, the graveyard is not *physically crowded*, because the circumference of its orbits exceeds 260,000 km, but the positions of the satellites are uncontrolled, and it is little short of a miracle that there have been no collisions.

The situation has arisen simply because of the number of satellites launched to GEO, and retired to the graveyard, coupled of course with the fact that no one has developed the technology to remove them. Around 2015, the satellites launched in the 1990s were reaching the ends of their operational lives, or had passed it, and because the 1990s had been a relatively busy decade for satellite systems a large number came up for renewal at around the same time. This had the knock-on effect of an equally large number being sent to the graveyard.

A similar replacement boom in the 2030s coincided with a boom in the telecom industry, during which the demand for satellites, as an integral part of that industry, rose beyond all expectations. More advanced compression techniques coupled with the development of large new tracts of frequency spectrum meant that satellites could, once again, fill gaps in terrestrial coverage caused by increasing demands from handheld communications devices. Thanks to those satellites and their successors, today's data, video, communications, and domestic security needs are met by a single handheld or wearable unit.

Although the 2030s batch of satellites had longer lifetimes than previous batches, the last few years have seen the initiation of yet another major renewal cycle and a further boost to the population of the graveyard orbits. The result is that, today, there are almost 1000 uncontrolled satellites of various vintages wandering around looking for something to hit.

Although no collisions have occurred, a fragmentation event last year in a lower graveyard orbit has put monitoring systems on high alert. It appears that a satellite boosted in the mid-2030s was not properly passivated on reaching the graveyard—in other words, a small amount of propellant or pressurant remained in its tanks—and an explosion occurred. Unfortunately, this proves the warnings made many years ago that graveyard satellites are little more than time bombs with randomized triggers.

The Space Surveillance Network has attempted to track the fragments produced by the explosion, but its facilities are overstretched and outdated as a result of funding cutbacks and a lack of international consensus on how to deal with the orbital debris problem. According to recent reports, the explosion produced several hundred trackable debris objects (and who knows how many nontrackable ones) that are currently swarming among the graveyard satellites. More than that, however, at least 100 objects, some the size of a football, are in orbits that intersect GEO itself. This means that operational satellites and all of the services they provide are also at risk.

Although it might not be politically correct to say "I told you so," it is fair to say that the space community has been aware of this possibility for at least the past 60 years. So why is the communications industry facing this dilemma today? It dates back to the Barcelona Summit of the early 2020s, when a major international meeting organized under the auspices of the United Nations failed to reach agreement on a set of proposals concerning, among other things, the control of the geostationary environment and its preservation as a commercial resource.

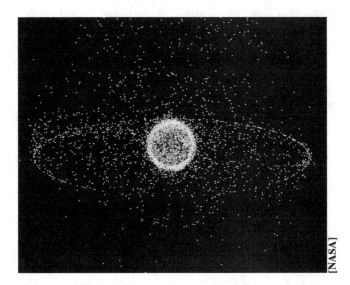

[NASA]

Fig. 89   Earth orbits are even more crowded in 2057 than they were at the turn of the century.

Continuing its philosophy of "freedom of access for any developers capable of establishing commercial services," the United States delegation vetoed all protection measures, criticizing them as "prohibition on an inter-planetary scale." It was felt at the time that, without U.S. support, any debris-mitigation measures instigated by the other nations would be unworkable, because the additional costs incurred by non-U.S. industry would place it at an economic disadvantage.

Although, in one sense, it is true that any international policy to which the United States is not a signatory is ineffectual—because it takes only one bad apple to contaminate the barrel (and one satellite explosion to contaminate an orbit)—developments have shown that the international space community was wrong to let the matter lie.

20/20 HINDSIGHT

It seems that, once again, the community has failed to learn from history. For example, who now remembers the star-crossed constellation of Ukrainian nanosats launched to a 900-km low Earth orbit in 2025? Despite fears that the 1028 10-kg, gravity-gradient stabilized spacecraft were tantamount to "designer debris," the constellation was deployed without sufficient consideration of its management, renewal strategy, and eventual decommissioning. Because the satellites carried no onboard propulsion system (apart from a deployable solar sail for orbital spacing adjustments), they had no energy source for deorbiting maneuvers.

Moreover, the misguided design decision to place them at an altitude of 900 km meant that natural clearing as a result of atmospheric friction could take upwards of 100 years, whereas had they been deployed at 600 km they would have reentered within about 15 years.

As it turned out, in 2027, less than two years after the constellation was completed, one of the nanosats hit an upper stage from one of the constella-tion's launch vehicles that had not been deorbited because of an electrical fault. As predicted by orbit evolution software, the resultant debris spread itself around the orbit, gradually impacting other objects and creating more debris. The much-feared "cascade effect," first mooted in the late 20th century, had finally occurred, with the result that the 900-km LEO (plus or minus 50 km for safety) will be unusable for a century or more.

Among other things, this affected UN plans for a disaster-monitoring constellation—using a much smaller number of satellites—which was to have occupied a similar orbit. The redesign costs and the difficulty in arrang-ing insurance coverage for such constellations put the project on hold indefinitely.

Again, hindsight shows that had an enforceable policy controlling the use of orbits been agreed earlier that decade, the disaster might have been avoided. Of course, if a manned station had been deployed at that altitude, the disaster would have been far worse . . . and possibly not forgotten.

In fact, there has been a precedent for such a disaster, which is now so far in the past that it ranks with the historic Space Shuttle accidents of 1986 and 2003. In 2009, China's first national space station, the *Feng Shui* (Fig. 90), was hit by debris, which ruptured a poorly shielded exterior propellant tank, causing an explosion. Its accommodation module suffered an explosive decompression, and two Chinese astronauts died—the first people to perish *in space*, as opposed to *in transit* to or from space.

As is the way with such fragmentation events, the story did not end there. Because the *Feng Shui* had been placed in an orbit of similar altitude to the International Space Station (ISS), to make in-orbit rescue missions easier, the debris from the explosion then threatened the ISS. Predictions for the evolution of the debris cloud showed that the ISS would suffer impacts within the following month, and it was decided to evacuate the station. This action alone was politically significant because it marked the first time since the ISS had become permanently manned that there were no human beings in space—not exactly a red letter day for space history!

Later analysis of the *Feng Shui* produced an ironic twist to the story . . . and another indication of how interdependent space activities can be.

Fig. 90   China's *Feng Shui* space station, which was hit by debris in 2009 (in reality, the ISS).

Discovered embedded in the module wall were the remains of an American torque wrench of the type used on the ISS (Fig. 91)—in fact the very one that a spacewalking astronaut had "dropped" two years earlier. In effect, he had set in motion a chain of events that would kill two Chinese astronauts and cause the first evacuation of the ISS.

Perhaps surprisingly, the "Feng Shui incident" failed to have a significant impact on the way the space community did business in orbit. Indeed, in some quarters it was seen as an example of the "sacrifice that had to be made in the greater name of space exploration and development." Following the investigation, China's response was to declare their fallen astronauts heroes of the state, deorbit the *Feng Shui*, and replace it with a larger, more capable space station. The shielding on the ISS was beefed up, and it was business as usual within 12 months of the incident.

When the ISS was finally decommissioned in 2020, a survey crew was dispatched to document the external condition of the station, and some components were removed and returned to Earth. It was found that in its two-decade lifetime the inhabited modules had suffered more than 1000 impacts of 5 mm or greater and hundreds of thousands of smaller ones. The solar-array panels, which increased the station's overall dimensions to about the size of a soccer pitch (Fig. 92), were so peppered with holes that one surveyor likened them to Swiss cheese. The largest holes exceeded 10 cm in diameter, affecting several solar cells, but there were many other

Fig. 91 A wrench dropped by an astronaut caused the *Feng Shui* disaster (in reality, an astronaut working at the ISS).

smaller ones that had knocked out individual cells. Were it not for the fact that arrays are formed from multiple parallel strings of cells, they would probably not have worked at all.

In a sense the damage came as no surprise, because spacecraft recovered in the late 20th century had shown similar damage, but this time the figures seemed worse because of the size of the spacecraft. However, it was because the space community thought it had "seen it all before" that the advice to the debris panel of the 2022 UN Outer Space Protection Conference was that "no further enhancement of debris mitigation measures was necessary."

Again, in hindsight, this has proved to be misguided. In fact, it has become clear that the conclusions were actively guided by those with political influence who wanted to limit the costs to governments and by corporations chiefly concerned with profits and shareholder values. It is also clear that the satellite industry has been content to exploit the resource of GEO, as long as it remains available, without providing the safeguards that would protect that resource for future generations of satellite users and operators.

EXPLOITING RESOURCES

The history of the Space Age has shown us that users of the natural resources of the space environment have, in general, followed the example of terrestrial developers: exploit and exhaust a resource, then move on to

Fig. 92   The ISS at completion—about the size of a soccer pitch.

the next. This has happened with low Earth orbit to the extent that there are now several distinct orbital zones no longer safe for manned stations and no longer insurable for unmanned spacecraft.

The space insurance industry had finally appeared to be on a rising trend by 2015, when the comsat boom coincided with the maturation of the Earth-imaging market. A record number of spacecraft were in need of insurance, and business was good for the insurance community. That, of course, was before an antisatellite weapon test, believed to have originated in one of the Arab states, went wrong. As a result of a command and control error, its delivery vehicle strayed into the orbit used by one of the leading Earth-imaging companies, locked onto one of its satellites, and destroyed it. Far from being a test gone wrong, some critical commentators considered it to be a perfect demonstration of the system's capabilities, the deliberate destruction of what some considered a spy satellite; others characterized it simply as an act of terrorism.

As a result of the test, the set of sun-synchronous orbits used by the Earth observation community became contaminated with debris, and imaging satellites became uninsurable. Several insurance syndicates, which had bet on the boom related to this application, went out of business, as did many corporations built on the back of the imaging business. Thousands of people in the space-related industries lost their jobs.

Meanwhile, those who had hoped to see Arthur C. Clarke's concept of a space elevator become fact were to be disappointed: even with the technology of carbon nanotubes, the thought of a structure linking Earth to geostationary orbit would remain untenable as long as the debris belts existed.

Over the next few years, the communications boom subsided as operators completed the replacement of their systems, and the insurance market, having dropped its rates too low to build up much of a reserve, was further decimated. The worry today, considering the overcrowding of the geostationary and graveyard orbits, is that an incident at these altitudes will make geostationary satellites uninsurable and the space insurance industry redundant.

EXPANDING THE ENVELOPE

But all this talk of Earth orbit is parochial. What about the enormous volume of space beyond the confines of Earth? As some would say, we might have made a mess of our immediate vicinity, but there's a whole universe out there.

Unfortunately, it is this philosophy that has driven the exploration agendas of the world's space agencies in the post space station era and capped the commercial development of space.

Although it is true that there are still space stations in LEO, they have not developed into the orbiting space hotels promised earlier in the Space Age. Because of the debris problem, LEO is simply too dangerous for tourists.

Adventure tourism is one thing, but adventure tourism without insurance is quite another matter.

Of course, it could have been different. Before the situation became as bad as it is today, the entrepreneurial sector made a brave attempt to open the space frontier to ordinary mortals (albeit those with plenty of disposable income). Well-off individuals had been visiting the ISS since 2001, when the first millionaire spent a week onboard, but space tourism did not really take off until 2015, after NASA had decommissioned its fleet of Space Shuttles and sold one to an independent operator.

That operator was finally able to do what tourism pundits had been predicting since the Shuttles were built: convert one of them into an orbital tour bus. Fifty lucky, rich, and relatively fearless individuals were strapped aboard each flight for a 10-orbit cruise past the ISS, which they were allowed to peer at from a safe distance. Unfortunately, as the flight rate increased, the Shuttle orbiter fast approached the end of its design life and began to show its age, both in terms of physical deterioration and downtime. It was close to 40 years old by the time it stopped flying, and the operator had not been able to build a replacement. This was mainly because of the difficulty in raising finances, which itself was a result of the perceived risks of manned spaceflight in LEO.

Space stations in LEO had once been seen as stepping stones to the Moon and the planets, a sort of orbiting pit stop and maintenance center, but when it became clear that a mass of debris shielding equivalent to that of the station itself would have to be lifted into orbit, the concept lost its attraction.

As a result, spacecraft were increasingly boosted on direct trajectories to their final destinations, ensuring that they spent the minimum time in the debris belts. LEO had effectively been bypassed. In cases where it was necessary to construct larger spacecraft in LEO, a very low orbit was chosen, and the job was done quickly to avoid reentry caused by atmospheric friction. Apart from a light rain of material from higher orbits, these orbits remain relatively free of debris.

Thus it was that, having abused and discarded low Earth orbit, explorers, developers, and entrepreneurs set their sights on the Moon.

RETURN TO THE MOON

The move to reconquer the Moon began with a dribble of unmanned flights in the first two decades of the century, including a number of relatively innocuous agency missions that failed to capture the public's attention. It was the soft landing of a fleet of teleoperated rovers funded by theme park operators that suddenly awakened people to the opportunities (Fig. 93). They allowed fee-paying members of the public to drive the rovers across the flat lunar seas, or maria, negotiating small rocks and hollows as they went.

Fig. 93   Spiderbot—innocent party in the field of remote tourism?

The novelty of navigating the barren lunar wastes soon wore off, however, and operators were forced to target their rovers at more sensitive sites. The obvious "big draw" was the site of the Apollo 11 landing of 1969, the place where men from Earth had first set foot on an alien world. Thus, in 2017, when the Apollotours-1 Rover touched down in the Sea of Tranquillity just 2 km from the Eagle lunar module, the eyes of the world were focused on their video screens.

The result had been predicted 30 or 40 years earlier. At first, Apollotours, Inc., acted responsibly, and the rover was limited to the periphery of the Apollo landing site, from where it used its camera's zoom lens to approach the Eagle's descent stage and scientific equipment. But the draw of the historic ladder which Armstrong and Aldrin had descended, and the plaque inscribed "they came in peace for all mankind," was too great to resist. Many of the astronauts' footprints, which had remained unchanged for almost 50 years, were obliterated by the rover's wheel tracks within a week. One careless driver bumped into the pole holding the famous flag, knocking it to a crazy angle . . . and the spell was broken (Fig. 94).

This was not what people wanted to see. The dream was to visit the Apollo landing site, however vicariously, and view it as it was a moment after the astronauts had left. The Apollotours rover had trashed the landing site as effectively as earlier hoards of unfettered tourists had trashed the Acropolis and Machu Pichu.

But this was as nothing compared to what Lunadevelopers, Inc., did to the crater Plato, one of the most popular and widely observed features on the lunar surface. An early Prospector spacecraft, launched in 2020, had

[Mark Williamson/NASA]

Fig. 94   Is mindless vandalism an inevitable adjunct to lunar tourism?

discovered a perplexing crystalline structure just below the crater floor and then broken a set of diamond-tipped drills attempting to gain a sample. Subsequent spectroscopic analysis of the material vacuumed from the hole had suggested that the structure itself was composed of a diamond-like substance. Although diamonds were still being mined on Earth, the prospect of being able to market lunar diamonds had an obvious appeal.

The rest, as they say, is history. Lunadevelopers organized a more complex Prospector mission and succeeded in confirming the discovery. The reaction of the financial industry was like that of a biologist discovering a new life-form. The stocks of Lunadevelopers, Inc., soared to unprecedented heights, and the first manned lunar mission of the 21st century was underway.

From the establishment of the first mining modules in 2035, to 2050 when the mine was abandoned, the company's scrapers stripped away the dark surface of the crater floor, constructing spoil heaps as they went. Within 10 years, they had scoured a grid-like pattern measuring 10 × 20 km in the center of the 100-km crater, and by 2050 two smaller sites had been similarly developed. To this day they are easily visible with only an amateur telescope because the gridlines and spoil heaps are bright compared with the undisturbed lava of the crater floor. What 17th century German astronomer Johannes Hevelius had called The Great Black Lake had become a great black scratch pad (Fig. 95).

The sad thing for Lunadevelopers—though some would call it justice— was the relatively short-lived fashion for lunar diamonds. As their novelty value receded, they were seen as overpriced—as well they might at 20 times that of their terrestrial counterparts. Lunadevelopers stock plummeted in the late 2040s, the company filed for bankruptcy, and its billion-dollar mining infrastructure was left sitting in the decimated crater. Then, last year, the propellant tank of a discarded lunar transporter exploded—the result of lack of maintenance and unremitting thermal stress, and debris was scattered over a wide area. It was clear to all that the passivation measures adopted for orbiting spacecraft had not been extended to lunar surface vehicles.

Nor does the international space community seem to have learned any-thing from the Lunadevelopers fiasco. The dozen or so other commercial development sites on the Moon are, today, troubled by little more than a basic set of guidelines on the treatment of the lunar environment, and lunar development remains a free-for-all.

[Mark Williamson]

Fig. 95   The result of strip-mining the crater Plato between 2035 and 2050.

This includes the development of a lunar hotel by MoonResorts, Inc., at Hadley Rille in the lunar Apennines, just 2 km from the Apollo 15 landing site. Although protesters were out in force at the launch of the initial ground-breaking mission, space agencies and other institutional bodies have remained silent on the environmental issues.

First reports of the sortie mission revealed it to be little more than a grand expedition to plunder the historic artifacts of Apollo 15. As a fund-raising exercise, the developers sold tickets—estimated at $500 million apiece—to a select cadre of billionaire collectors, who joined MoonResorts' engineers on the sortie. Allegedly, though tight security made it impossible to verify, the collectors were each allowed to return with up to 5 kg of "salvaged" Apollo hardware and another 10 kg of lunar materials (either for their private collections or for resale).

MoonResorts, itself, is known to have salvaged parts of the ALSEP experiments package, and, most significantly, the Apollo 15 lunar roving vehicle (the first vehicle ever driven by a man on the surface of another planetary

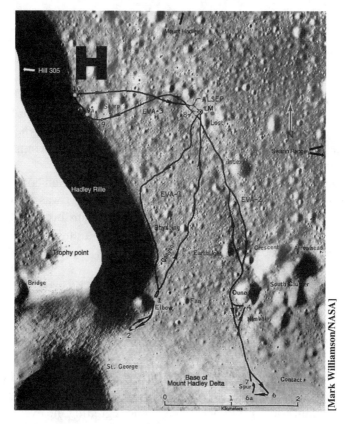

[Mark Williamson/NASA]

Fig. 96   The site of the *MoonResorts Lunar Hotel* at Hadley Rille.

Fig. 97   Will this Apollo Lunar Roving Vehicle be auctioned to the highest bidder?

body). Word is that the LRV is to be auctioned privately in the near future for an estimated sale price as high as $10 billion (Fig. 97). And what will the tourists see when they get there? A mock-up of the LRV, complete with two space-suited figures, posed by MoonResorts' staff.

The development has been criticized as the "Disneyfication" of the Moon, but the term does a deep injustice to Mr. Disney, who was never in the business of plundering archaeological sites before transforming them into theme parks. One wonders what developers will do on Mars when their clientele become bored with the Moon.

CELEBRATING A CENTURY OF SPACE TRAVEL

So how are we celebrating 100 years of the Space Age? Is anyone celebrating?

The answer, of course, is yes, because the passing of 100 years of anything or anyone is always worth celebrating. But we are celebrating 100 years of the Space Age in the same way that we celebrate the end of a war: pleased that we have come out of it in one piece, while regretting the events that led to the celebration.

Chief among the regrets are those of the space industry, which had hoped to do so much more in 100 years. As a result of a lack of protection afforded to the low-altitude orbits, Earth imaging and space tourism, in turn, failed to supersede satellite communications as the most successful space application. And, today, that application itself is under threat.

For manned applications, the increasing risks of travel to and accommodation in LEO have limited the development of research stations, engineering platforms, and tourist facilities. And this, in turn, has limited the development of low-cost vehicles to transport people and cargo to LEO. Companies that would have developed this infrastructure have either gone out of business, or not been formed in the first place.

Although attempts are still being made to develop the Moon as a materials resource and tourist destination, developments are tentative and tainted with short-sightedness and greed. Although the freedom to explore and develop an environment provides short-term gains for the few, it rarely results in long-term gains for the many. It is an indictment of government and commerce that, in 100 years of space exploration and development, this lesson has still to be learned.

For those who had hoped to colonize the planets, deliver unlimited solar power from orbiting stations, and protect the Earth's population from the seemingly inevitable meteor impact, the centenary of the Space Age might well be spent crying in one's beer.

The space protection lobby might be forgiven a certain smugness in the light of the situation, but it gives them no pleasure to be proved correct. Many of them had jobs in the space industry and saw those jobs become redundant as the need for their professional skills diminished.

Today, we stand on a threshold to space, not knowing what lies ahead, as our forebears did 100 years ago. But it is a different threshold to that of 1957: it is littered with the garbage of those who have used and abused the space environment before us. Our options are fewer, and our resolve is sullied by the failures of the past. What was once the final frontier, in the optimistic sense of the last hurdle in the race, has become a formidable barrier. It might be another 100 years before we are able to breech that barrier and continue our journey in the universe.

Chapter 8

Scenario B: Control and Confine

Chapters 7, 8, and 9 are written from the perspective of the year 2057, the 100th anniversary of the launch of Sputnik 1 and the centenary of the Space Age. They offer three different views of how the future might turn out for space exploration and development under certain premises.

This second scenario assumes that strict policies of control and protection are instigated and enforced.

Keynote Speech to the 108th International Astronautical Congress
Baghdad International Conference Center, Iraq
October 2057

Mr. President, honored guests, and delegates, it is October 2057, and the Space Age is 100 years of age. A centenary such as this should be worthy of extensive celebration, at least among the space professionals gathered here today, but is it really worth raising a glass for?

In 1957, for those of our forebears with sufficient foresight, the potential for space exploration and development seemed as infinite as the universe itself. There were countless discoveries to be made, countless new worlds to conquer, and countless ways to make money from what we now call the space environment.

Today, however, there are many who would argue that the human race has failed to recognize that potential and, by constructing an almost impenetrable barrier of rules and regulations, has restricted its opportunities and quashed its entrepreneurial spirit.

So, how did we get to where we are today?

REGULATION AND LEGISLATION

Although the problems of orbital debris and their possible solutions were discussed at length in the late 20th century, at meetings such as the International Astronautical Congress, the debris mitigation regulations that exist today owe their foundation to the first decade of the 21st century. In fact, the spark that ignited the founding discussions can be traced to a single event: the devastating impact suffered by the *Feng Shui*, China's first space station, in 2009.

Once the international space community had absorbed the fact that the impact had caused the first deaths in orbit, and had *been caused* by a wrench released from the International Space Station, it began to work to avoid a repeat of the disaster.

In the following year, a number of preparatory meetings were held under the auspices of the United Nations Committee on the Peaceful Uses of Outer Space (UNCOPUOS), with the intention of gaining international support for the formation of an international body for space debris mitigation. It soon became clear, however, that confining the remit of the body to orbital debris was insufficient, given that orbital space was but one of the natural resources of the space environment under threat.

As a result, UNCOPUOS extended its hearings into 2011 to allow nongovernmental organisations, or NGOs, to present their findings and opinions to the review panel. For example, a body representing the satellite industry pointed out that any regulations formulated would have to apply equally to manufacturers of all nations; otherwise, the cost of additional miti-gation measures borne by signatory nations would place their manufacturers at a competitive disadvantage. Similar pleas were made by launch provider and tourism industry organizations, amid many other calls for a "level playing field."

In addition, a delegation from the International Space Environment Protection Union (ISEPU) spearheaded its campaign with a forceful position paper on the need to extend protection to the planetary bodies of the solar system, including not only the planets themselves, but also their natural satellites and major asteroids and comets. It was supported by a number of scientific bodies which feared that uncontrolled exploration and exploitation of the space environment would lead to the contamination of planetary bodies and, possibly, the elimination of indigenous life-forms.

As a result, it was decided at an international level that the new regulatory environment should be supported by legislation if it were to be effective and enforceable, and the International Institute for Space Law (IISL) was called upon to formulate a plan to extend the existing body of space law. This led, over the next few years, to an extension of the 1972 Liability Convention, concerning compensation issues, and the 1976 Registration Convention, to include the obligatory publication of plans to explore or develop planetary sites.

Finally, the 1967 Outer Space Treaty was significantly revised to include measures for protection of the space environment in its widest terms. This action finally laid to rest the disagreement over the 1984 Moon Agreement, which had attracted only a handful of signatories, because the Revised Outer Space Treaty (ROST) of 2020 effectively superseded its provisions.

However, even before the 2020 ROST was enacted, the United Nations had secured international agreement to a space environment protection policy that encompassed orbits around the Earth and the Moon, and the surfaces of the Moon, Mars, and Earth-crossing asteroids. In 2016, ISEPU was upgraded to UN committee status, and the International Telecommunication Union (ITU) was incorporated under the same umbrella. For the first time since the ITU was formed in 1934, it was endowed with the powers to enforce its decisions, and the formerly chaotic system of orbital and frequency resource allocation became ordered and predictable.

ENFORCEMENT REGIME

The first international plenary meeting of UNISEPU, held in Geneva in September 2016, was a *tour de force* in space policy formation, and because it was timed to precede the 67th International Astronautical Congress, also in Geneva, it offered an extended forum for international deliberation.

The primary result was an international agreement on the protection of orbital and radio-frequency resources, which stipulated a raft of debris-mitigation measures to ensure that all space objects—whether spacecraft or launch vehicles—added nothing to the existing debris population. A second measure covered the surface environments of the planetary bodies, making any form of contamination illegal under international law and providing for cleanup measures in case of unavoidable accidents.

As far as enforcement was concerned, any organization proposing to explore or develop any part of the space environment was obliged to publish and agree an environmental protection plan with UNISEPU and show that it had the financial resources not only to conduct the mission in accordance with the regulations, but also to initiate a recovery, cleanup, and compensation plan if the mission failed to meet them. The financial penalties were stiff, and the regulations threatened custodial sentences for the worst cases of infringement.

The enforcement regime did not stop there, however. As part of its grand plan, UNISEPU and its sister organizations initiated the Space Control Program—in effect, a space traffic control system analogous to air traffic control on Earth. The U.S.-based Space Surveillance Network (SSN) was expanded on a global level to enable controllers to monitor just about anything that moved in low Earth orbit and beyond. All satellites and other spacecraft were allocated a volume of space, a *cordon sanitaire*, within which no other space object was allowed; and if the cordon was breached, there would be financially punitive repercussions. Manned space stations were allocated the largest cordons, which were strictly controlled to manage the approach and docking of supply vehicles and the like.

[EADS]

Fig. 98   An Orbital Enforcement Vehicle operated on behalf of UNISEPU by a private contractor (in reality, ESA's ATV at the ISS).

The enforcement measures of the Space Control Program were applied with draconian enthusiasm.

By 2025, a version of the orbital delivery vehicle first fielded by NASA in 2012 had been developed as an orbital enforcement vehicle (OEV) and was being operated on behalf of UNISEPU by a private contractor (Fig. 98). One OEV was stationed at the ISS and another in polar orbit at a Japanese polar platform. Their role was to investigate transgressions of the international agreements and even forcibly deorbit spacecraft that failed to meet requirements. A similar deorbiter, albeit unmanned, had been despatched to geostationary orbit the previous year, and by 2030 a fleet of these spacecraft was systematically clearing the graveyard orbits.

For those of you who have difficulty recalling the dates, the table should serve as a reminder.

RESULTS

The protection agreements brought about many changes in the way space systems were operated, most of them positive. Debris-mitigation measures helped to keep the most popular orbits relatively clear, and the dreaded "cascade effect" was avoided. The international upgrade of the SSN introduced a new era of data sharing and allowed the Space Control Program to have the desired effect, in that significant in-orbit collisions became a distant memory. And the moratorium on in-orbit military tests was upheld,

Table 6   Key dates

Date	Event
2009	*Feng Shui* incident
2010	UNCOPUOS preparatory meetings
2011	Extended hearings including ISEPU
2011–2014	IISL plans extensions to space treaties
2016	ISEPU becomes UNISEPU and holds plenary in Geneva
2018–2020	Space Control Program initiated; SNN expanded
2020	ROST enacted
2025	OEV developed and launched
2030	OEV fleet in graveyard orbit
2045	Graveyard mostly cleared

with the result that, apart from remote surveillance, communications, and navigation applications, space was effectively demilitarized.

Meanwhile, supported by military and government contracts, the Earth-imaging industry began to take off, and by 2020 commercial imaging had replaced communications as the most profitable commercial space application. In parallel, stricter controls of the orbital environment reduced damage claims on insurance policies, and the space insurance industry attained a previously unheard of level of stability.

To further protect LEO, the deployment of uncontrolled constellations of nanosats and picosats was outlawed, and all space objects, of any size or mass, could only be launched once a comprehensive deorbiting plan had been approved. Unmanned deorbiters were deployed in all of the major orbits, and by 2045, as the table shows, much of the geostationary graveyard had been cleared; indeed, the market for replacement deorbiters had begun to challenge that for the satellites themselves (Fig. 99).

So many improvements had been made as a result of the agreements, and so well had the benefits been sold to the space community, that it appeared no wrong could come of the new regime. At first, even the negative mindset of those in the industry seemed to improve.

But all was not well, as we have come to learn to our cost. The problem was that the space community's attention had been distracted from its primary task—the exploration and development of space—not intentionally, but as a result of its focus on other tasks.

Certainly, the pace of constitutional development following the *Feng Shui* incident had been breakneck, almost frenzied compared with previous schedules, and there is little doubt that the second decade of the 21st century will be cited by future historians as a period of unprecedented international cooperation. For a while, it seemed that the political world

Fig. 99   An unmanned spacecraft deorbiter (in reality, a NASA artist's concept for an Orbital Maneuvering Vehicle from 1984).

had discovered a common duty of care to which all could subscribe. Space belonged to none of the Earth's governments, but was potentially a resource for all. The attraction of being seen to preserve an environment that had a bearing on the whole planet, without actually surrendering anything of their own, was too great for politicians to ignore.

Unfortunately, governmental benevolence did not extend to patronage, and the space industry was obliged to fund improvements to its technology on a commercial basis. There were some space agency technology contracts to be won, of course, but debris-mitigation measures were funded largely by satellite and launch vehicle-purchasers, who were obliged to pay higher prices for the "improved" hardware.

But it was not hardware costs alone that made satellites and rockets more expensive; it was the bureaucracy that accompanied the new regime. Space programs had always been top heavy as far as paperwork was concerned, but the new regulations required an additional level of technical accountability and expensive additional testing. Having honed their development schedules to reduce delivery times, manufacturers were once again forced to increase them.

The result was that space systems suppliers found it increasingly difficult to make a profit and, having already undergone a series of mergers and purges in the early years of the century, were forced to make additional cutbacks. The employment situation improved slightly around 2015, when a large number of geostationary communications satellites became due for replacement, but it was a short-lived boom. The industry was left with one major satellite contractor in the United States, one in Europe, and a couple of smaller ones in the Far East, while a similar contraction affected the launch industry.

The additional costs of the bureaucracy, and the potential fines awaiting those who broke the rules, persuaded most of the fledging space nations to abandon all hopes of joining the wider community. Space was back where it started: in the hands of the superpowers.

Perhaps more importantly, the regulation regime eliminated most of the commercial entrepreneurs, who preferred to invest in less risky ventures nearer to home. The most important result of this withdrawal is the current lack of development in low-cost access to space and all that that implies.

SPACE FOR THE PEOPLE?

When the century was young, it was felt by many that the key to opening up the space frontier, which really meant making money from space, was space tourism. The ailing Russian space industry was charging $20 million a seat on its Soyuz ferries to the ISS, and, around the world, entrepreneurs were competing to win a competition to build a spacecraft that would carry ordinary mortals on a suborbital trip of a lifetime. Progress was slow, but at least interest was on an upward curve.

However, just as China was busy making plans for tourists to visit the *Feng Shui*, the nation suffered the loss of its first space station, two of its astronauts, and a good deal of pride. Having been forced to deorbit the *Feng Shui* because its structural integrity had been compromised, China wanted to replace it. However, like Russia before it, China found that funds for space were in short supply in the face of its second great cultural revolution: improved political openness, increasing international cooperation, and freedom of speech for its population had uncomfortable side effects for the nation's space aspirations.

Despite a loss of face, China's increasingly cooperative spirit made it a key player in UNISEPU, and later agreements led to the launch of a joint Chinese–European station dedicated to Earth observation and surveillance (Fig. 100).

Fig. 100   The joint Chinese–European space station (in reality, ESA's Columbus module).

However, the *Feng Shui* incident had ramifications far beyond China. NASA and the other ISS partners had decided to evacuate the International Space Station in case it was seriously damaged by debris from the *Feng Shui*, and a year passed before the respective agencies overcame their apprehension. Led by NASA, as fearful of transporting civilians to the ISS as it had been flying them on the Shuttle following the Challenger accident, the agencies placed a moratorium on tourist visits.

When NASA came to decommission its first Shuttle in 2010, it was bought by an independent operator, but that operator's business plan involved station supply and debris removal activities, not tourism. There was, after all, nowhere to go, and the *cordon sanitaire* rules for existing space stations meant that rubber-necking tourists were, to say the least, unwelcome.

Nevertheless, when the ISS itself was decommissioned in 2020 and the construction of ISS-Beta was almost complete, the agencies succumbed to demand from the nascent space tourism industry to lease parts of the ISS to a commercial contractor. So, following some refitting and refurbishment, the world's first space hotel was open for business. The orbital cleanup campaign had secured the future of a new space application: LEO was considered safe enough for tourists, travel insurance was available, and space vehicles could be designed without the need for massive, and prohibitively expensive, shielding.

The downside of the space protection legislation was that manned activities had been largely confined to Earth orbit, with the result that the space community had become Earth-serving and inward looking. The Moon had been designated a UN World Heritage Site, which meant that it was protected from commercial and military development and, like Antarctica, was intended to remain a haven for scientific investigation.

At first this seemed ideal for lunar science and, in a belated response to the U.S. presidential decree of 2004, exciting plans were made to return mankind to the Moon and send the first astronauts to Mars. Unfortunately, there was insufficient attention to history, and political detractors gained an important foothold: the new program was "little more than Apollo all over again," they said, and people started to listen.

The crux of the problem was that, with industrial development banned, industry had little interest in the Moon. Of course, it would have been glad to build spacecraft and bases in support of scientific investigation, but without the means for income generation there was insufficient money in agency budgets, and too little political incentive, to return to our nearest planetary neighbour. Having spent time and money making LEO habitable, the agencies were focused on operations and applications closer to Earth. Some of the detractors had been proved right: the ISS and its successor were not stepping stones to the Moon and Mars; they were islands in a stream that led nowhere.

Early in the century, the exploration of the Moon had been confined to a few soft landers and simple rovers. The early rovers were of limited success, however, because the hardware was designed and built to a limited budget for theme park operators. In most cases, before the operators had recouped their investment, the rovers stopped working because of dust contamination, heat stroke, or simply because they had been driven into a crater from which they could not escape. Having responded to the publicity and visited the theme parks in their droves, prospective lunar drivers were disappointed to find "exhibit under renovation" notices. Once the protection legislation took hold, the cost of the spacecraft made remote tourism commercially unviable, and teleoperated rovers were dropped as quickly as they had been embraced.

Since 2050, hopes have arisen that some sort of nationalized lunar tourism might be organized, and contractors are currently bidding for contracts to build and operate a lunar visitor center near the Apollo 11 landing site. Based on the model of the national park on Earth, the Tranquillity International Park would welcome visitors on carefully escorted—and highly regulated—tours that would include a stay in a lunar "resort hotel." The hotel and support facilities would be built underground and landscaped to maintain the appearance, as far as possible, of an undisturbed lunar sea. The Apollo exploration site itself would be enclosed in a large transparent dome adjacent to a museum, and the Apollo hardware would be protected from visitors by a specially designed conformal Perspex cladding, as would selected areas of the historic lunar surface.

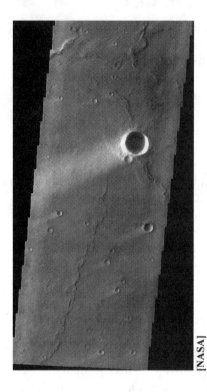

Fig. 101 Without planetary landers and rovers, any form of extraterrestrial life will remain undiscovered (image shows the interaction between wind and landforms on Mars).

No one really knows whether the program will go ahead because it relies on a large amount of funding being made available. There are calls to fully commercialize the program to allow the operator to make a profit from future lunar visitors, but the legislation in its current form forbids this. Apart from that, the business risk associated with such a speculative venture would seem too great to contemplate; the space industry itself has said on many occasions that it has no interest in museum development.

And what of scientific exploration beyond the Moon? While carefully designed orbiters continue to probe the secrets of the planets from their remote vantage points, the costs of complying with planetary protection regulations have made landers and rovers a difficult option. With the ever-present threat of international legal action, and punitive fines, should anything go wrong, agencies and laboratories have found little incentive in planetary exploration.

The irony of the situation is that without planetary landers and rovers, any form of life that can exist on or beneath the surfaces of the planetary bodies, or indeed in their atmospheres, is likely to remain undiscovered. And if it remains undiscovered, we shall never know whether it was worth protecting (Fig. 101).

DISSENT RISING

Perhaps the biggest threat to the space environment, however, is the rising dissent within the United States where, despite the spirit of agreement of the last few decades, free enterprise and unilateralism still rule. Based no doubt of their reading of the nation's history of the last century, some of the more vocal politicians have begun to promulgate the slogan "prohibition doesn't work." The fear on behalf of the rest of the world is that America will withdraw from the space protection treaties, as it did from the weapons test ban treaty.

We shall then be back to the free-for-all situation that existed in the late 20th century, when individual nations did more or less as they wished in space. The protocols and legislation developed over the past half-century will be replaced by a form of space anarchy in which protection of the space environment will be subsumed or ignored.

Although America's unilateral withdrawal would undoubtedly be bad for international peace and cooperation, it is debatable whether it would actually be bad for space exploration and development. As it is, under the present protectionist regime, the space science community is constrained to mediocrity, the space industry is a shadow of its former self, and true commercial enterprise has been confined to Earth.

If our freedom of access to the space environment continues to be constrained in this manner, mankind will remain a single-planet species . . . and will one day be obliterated by that long-predicted "planet-busting" impact. Ironically, we shall have internationally based, highly reliable, and multiply redundant tracking systems to warn us of the event, but no in-orbit infrastructure to prevent it.

Ladies and gentlemen, it is customary in speeches of this type to conclude by looking towards the future for the space profession, hopefully in an optimistic vein. Sadly, I must fail you in this regard because if space exploration and development continues to be constrained in the way I have described, we simply won't have one.

Thank you for your attention.

Chapter 9

SCENARIO C: BALANCE AND BENEFIT

Chapters 7, 8, and 9 are written from the perspective of the year 2057, the 100th anniversary of the launch of Sputnik 1 and the centenary of the Space Age. They offer three different views of how the future might turn out for space exploration and development under certain premises.

This third scenario assumes a middle way that balances the needs for development and protection.

Chryse Planitia Station
Mars
4 October 2057
[*INN* newsfeed from Mars via interplanetary communications grid to Luna Intranet]

When the sun rose over Chryse Planitia this morning, it glinted on the surface of a geodesic dome that technicians had erected over the Viking 1 Historic Landing Site. The golden light, hitting the golden dome on what is otherwise known as the Gold Plain, provided a spectacular beginning to the day that marks the centenary of the Space Age. It seemed a fitting tribute to what some are calling the Golden Age of space exploration.

What better time to review the events that have brought space exploration and development to where they are today? A 100-year review of anything is no easy task, but it can help to show us where we came from and can even provide some pointers for the future.

OLD WAYS

Asked to identify the key factor that has encouraged space activity in the past 50 years, the Director of the Chryse Planitia research station, Dr. Raymond Crater, said simply: "sustainability." "On Earth, we had sustainable programs for forestry, fishing, and even the beginnings of sustainable energy production," he explained. "By contrast, the first 50 years of the Space Age were notable for their nonsustainable approach to exploration and development."

[NASA]

Fig. 102   The Viking 1 Historic Landing Site on Chryse Planitia before the dome was built.

Although it is hard to believe from today's perspective, the vast majority of space exploration missions were one-off events, operated by individual nations, and conducted without planning, or funding, the subsequent stages. Amazingly, the majority of space systems were new developments, rather than designs evolved from previously qualified hardware. According to Crater, "control systems were nonstandardized, incompatible and relied on 'heritage components,' which often became obsolete during the systems' operational lifetimes."

NASA's first space shuttle fleet is a prime example: it was an entirely new development in no way based on previous U.S. manned spacecraft, which were nonreusable reentry capsules, and once the fleet was disbanded its design was relegated to the history books. Moreover, because the design was based on 1970s technology and the fleet remained operational for some three decades, the orbiters were continually taken out of service for upgrades, not the least to update archaic onboard computers.

Another heavily criticized part of the early orbital infrastructure was the International Space Station (ISS), which, according to space activist Edward Gresshatch, "balanced on an operational knife-edge throughout its life." From its initial construction phases to its early in-orbit operation, it was continually threatened by problems with the Space Shuttle (the only means of module delivery); a poor resupply and crew rescue strategy; an unconvincing mission plan; and wavering international support (both moral and financial). And as the station aged, more and more crew time was spent on housekeeping and maintenance at the expense of science, especially after the reduction in crew size following the Columbia accident.

Not only were the space developments of the first 50 years unsustainable, says Gresshatch, they were "undertaken with insufficient regard to the inherent value of the resource being exploited."

Whether the resource was one of orbital space or planetary surface, there was a general attitude of ignorance concerning the fragility of the space environment. Had these attitudes been allowed to continue, the development of space would have been curtailed almost before it had begun, and we would not have reaped the benefits—financial, social, and cultural—that we have today.

FOUNDATION FOR SUCCESS

The sustainable approach had its roots in a precursor to the United Nations Space Council (UNSC) known as the International Space Environment Protection Union (ISEPU), a body formed in 2010. Its far-sighted members drafted the policies and legislation that promoted and encouraged "joined-up thinking" among bodies involved in both scientific exploration and commercial development. Although initially they met with widespread opposition to their views, the solidity of their commitment to space activity—and let us face it, the undesirability of the alternative—gave them the strength to continue.

The major irony, according to Raymond Crater, is that those who vehemently opposed the "protectionists" were the very same people who were desperate to develop and industrialize space and frustrated that it was not happening in the way they thought it should. What ISEPU did, in arguing that the space environment was special and worthy of protection, was to highlight its value, says Crater. And it was this appreciation of value that appealed to those who held the keys to space development: the financiers.

Although many members of ISEPU believed that space has a value in its own right and would often cite the uniqueness of its physical features and the sanctity of alien life (should it exist), they also had a pragmatic and long-term view of its value as an asset. In the words of space analyst, Rosemary Enscale, "the space resources of orbit space, frequency space, microgravity and low gravity environments, as well as surface real estate, were so beneficial to mankind that they could not bear to see their value reduced or degraded."

The challenge, she says, was to prove to the would-be developers that, by analogy with terrestrial developments, adoption of the sustainable and environmentally friendly model was in their long-term best interests. Although there were space laws to amend and ethical codes to agree along the way, the thoughtful, pragmatic, and multilateral methods employed allowed developers a sense of ownership in the regulatory procedures. And with this came the realization that their interests were being protected as much as those of the protectionists. A workable balance between development and protection had been struck.

BENEFITS

So let us look at what the sustainable approach has given us today, a century after Sputnik 1 broke the surly bonds of Earth to become its first artificial satellite.

First of all, we have a permanent orbiting infrastructure in low Earth orbit, polar orbit, and geostationary orbit, comprising both automated space platforms and manned stations. This is primarily because debris-mitigation policies designed by ISEPU were backed at government level and enacted on an international basis by all space-faring nations. Put simply, our orbital environment in now sufficiently safe for us to risk our lives and property there.

To maintain the safety of that environment, all satellites are now built with integral deorbiting systems, based either on ion thrusters or tether systems, dependent on their initial altitude. This means that once a spacecraft reaches the end of its operational life it can remove itself from orbit or decrease its altitude to rendezvous with an autonomous space tug, which will complete the process on its behalf. The additional expense of this orbital regeneration is now recognized as an inherent cost of doing business in space.

Similar technology has been brought to bear on the burgeoning population of defunct satellites in the geostationary graveyard orbit, while the amount of hardware launched to GEO in the first place has been reduced by the introduction of "plug-and-play" satellite platforms, which allow payloads to be swapped or reconfigured and platforms themselves to be refueled.

The decrease in orbital debris has also benefited the development of manned space stations, which are operated in a far more sustainable manner than the ISS. Following the success of the first Chinese station, which was decommissioned in 2015, China joined an international station development group, which began constructing the Gateway-1 station in 2020. As an indication of its permanence and utility, it was the first artificial-gravity station in LEO, a large rotating wheel of the type favored by 20th century science-fiction films (Fig. 103). As Edward Gresshatch points out, the fact that anyone could visit a space station without losing muscle mass and suffering bone demineralization made space "more accessible to ordinary folk," and, suddenly, the floodgates of space tourism opened.

Once it became clear that a sustainable infrastructure could be developed, funding became available for a more reliable and cost-effective means of access to be developed. As a result, it is possible today to launch a space transport on demand—if necessary within minutes—rather than wait

[NASA]

Fig. 103 An international station development group began constructing the Gateway-1 station in 2020.

weeks for a shuttle to be readied. In the same way that the commercial airlines opened up the world, says Gresshatch, "the commercial spacelines have unlocked Earth orbit." Space tourism is no longer restricted to multimillionaires; it is something that any committed and reasonably well-heeled adventure tourist can afford.

According to space scientist Dr. Perry G. Burns, a welcome spin-off from the development of the spinning space station was, rather paradoxically, the rise in microgravity applications. "Once transportation costs fell," explains Burns, "it became possible to launch and supply independent microgravity platforms that industrial entrepreneurs could use as a common business asset, leasing orbital modules in the same way they rent lab-space in business parks on Earth."

A more recent development was that of the Senior Orbital Community (SOC) Program, which welcomes the older generation to its "orbital refuge": "unencumbered by gravity, regain the mobility of youth" boasts its promotional material. The business case is based on the relative wealth of this sector of society, whose retirement packages provide a higher disposable income than any other consumer group, according to an SOC spokesman.

"The key to these and other developments," says Ed Gresshatch, "was the affordable ticket to space provided by the new generation of spaceliners." Of course, the transition from government-owned space shuttle to commercial passenger carrier was a sometimes painful process, punctuated by hardware failures and the occasional accident, but it was empowered by the new model of sustainability in the space community. "Once the X-Prize had been won and the U.S. Government had been persuaded to encourage commercial space ventures, a new space race was on," concludes Gresshatch.

Rose Enscale has studied the history behind these developments. "Now that we have such easy access to low Earth orbit, it has become as much a holiday destination as Hawaii was in the late 20th century," says Enscale. "Like the airliners before them, we take space transportation systems very much for granted. And, as with aircraft, we tend to forget that they have other uses."

Indeed, even before the first hypersonic passenger transports were certified for passenger travel, they were used for intercontinental mail delivery, carrying high-value packages from one side of the world to the other in a matter of hours. Among the first users to risk the new delivery system were electronics companies, desperate to deliver components to stalled production lines. Later, in cases where conventional aircraft would deliver too late, human organs for transplantation were added to the manifest. Today, of course, suborbital courier services are a recognized delivery option (Fig. 104).

[EADS]

Fig. 104 A hypersonic vehicle for sub-orbital mail delivery and troop deployment.

And once the suborbital vehicles had been man-rated, the military welcomed the technology with open arms. At last they had the means to deliver troops to a conflict zone, almost anywhere in the world, within an hour of launch. Since then, there have been many occasions when an early response helped prevent the escalation of military action and significant loss of life.

The revolution in space access has also brought about the realization of the dream of delivering power from space because it is now cost effective to transport the significant mass of the solar power satellite to orbit. Current figures show that, with new high-efficiency solar cells, the generation capacity of space solar power outweighs other renewable energy sources by a factor of three to one. The establishment of this space power infrastructure was unthinkable when access was expensive and debris was an ever-present threat to the hectares of solar arrays deployed by each satellite.

Debris also threatened the concept of the space elevator, which became at least theoretically feasible in the early part of this century following the development of carbon nanotube structures. Next year's inauguration of the first commercial elevator system should finally cast all doubts aside.

All this, and more, is a direct result of the policies of environmental protection pioneered by ISEPU in the first three decades of the century, policies that have even helped to boost the wider economy.

ECONOMIC BOOM

Having learned the lessons of short-termism, practiced particularly around the turn of the century, the world's financial institutions discovered (with a little help from ISEPU) that space offered the ultimate in long-term investment, while boosting the global economy.

During the boom of the 2030s, employment in the space industry quadrupled to meet the need for infrastructure, while according to financial analyst Diane Plexer, "company share prices reached high but sustainable levels and financing became available for new projects that met the sustainability criteria."

Significantly, the boom was not confined to a few leading nations, because with cheaper access to space—even the smaller nations could afford to develop, launch, and operate space systems. "This gave the smaller economies an enhanced feeling of ownership," explains Plexer, "and a desire to have a greater say in how the assets of the space environment should be used." As a result, increasing numbers of nations sent delegates to the UN Space Council, both broadening its constituency and endowing it with greater muscle within the international community. Space was no longer on the political sidelines; it was center stage.

Of course, there were still the usual ups and downs in the global economy, but rather than bumping along the bottom, it was riding a wave of relative prosperity. This meant that funding was available for even the most speculative space-related ventures.

This was shown to effect most recently when Integrated International, Inc. (I^3) opened its first lunar foundry for optical memory devices. Research had shown not only that the glassy substances required as a raw material for the devices were plentiful in the lunar regolith, but also that the one-sixth gravity was beneficial to the purification process. As Integrated's CEO, Kevin Larpanel, is keen to point out, "on-site processing reduces the mass that must be transported to the I^3 fabrication facility in LEO and significantly reduces operational costs."

Naturally, he adds, "the required environmental policies have been adhered to in the construction of the foundry." Regolith disturbed by strip mining is recompacted as it leaves the stripper, and a dusting vehicle that follows the stripper returns the surface to its former texture and color, rendering the strip invisible. The facility itself is buried underground, thus providing a natural radiation shield, and the surface is landscaped to mimic a natural appearance. To the casual observer, the only indications of a manned base are the adjacent spaceport and the surface vehicles.

The same standards of environmental awareness and sustainability have been applied to several other lunar projects, from far-side radio observatories to the development of world heritage areas surrounding early exploration sites, particularly the Apollo sites. In addition, by analogy with terrestrial practice, the UN Space Council has also designated Sites of Special Scientific Interest, Areas of Outstanding Natural Beauty, and International Parks, within which industrial development is not allowed (Fig. 105).

[NASA]

Fig. 105   Tranquillity Base International Park (in reality, the Apollo 11 landing site with Neil Armstrong working in the shadow of the lunar module).

Reginald O'Lith, leader of the lunar environmental lobby—popularly known as "The Greys"—is happy with the result. "The balance achieved has preserved important areas of the Moon for the enjoyment of and study by future generations," he says, "while allowing carefully controlled development in most other areas." And the compromise appears to have worked.

Moreover, according to Princeton University's communications analyst Professor R. F. Link, "the establishment of a cislunar trade route, complete with lunar orbit communications infrastructure, tracking stations and landing facilities, has made it economically viable to develop a lunar tourism industry." As he confirms, the first lunar hotel, built in 2045, has proved a great success, and others are planned. And following the acceptance of the 100-m microgravity relay by the International Olympic Committee, an event that owes its origins to the swimming pool at the Hilton Orbital Hotel, there are plans for a multitrack sports stadium on the Moon!

STEPPING STONES

Orbital operations and lunar development have been successful not only in their own right, but also in providing stepping stones to the planets.

The launch and orbital infrastructure that has been developed over the past few decades has brought about a confidence in orbital operations that was nonexistent before the debris-mitigation policies were enacted (Fig. 106). As a result, the permanent presence of people in space is no longer something special, as it was when the first international space station was under construction; it is simply a fact of life.

[EADS]

Fig. 106   An Earth-to-orbit crew transporter of the 2050s.

Likewise, it is hard to believe that almost 50 years separated the departure of the final Apollo astronauts in 1972 and the landing of the first Lunabase astronauts in 2020, especially when one considers what has been achieved in the subsequent 37 years. As Chryse Station's Ray Crater confirms, the lunar bases which followed that early field trip have proved to be an excellent grounding for the phase of Mars exploration currently underway.

It is also hard to believe that, at the turn of the century, manned Mars missions were seen as impractical and risky simply because of the travel time. With mission durations in excess of two years, this was understandable, but now that Mars is little more than a month away the goalposts have been well and truly shifted. Of course, we have the technology of nuclear-electric propulsion to thank for that, the same technology that will be used to deflect the fabled "Earth-destroying asteroid" when it appears on our screens.

It is the technology, and the confidence in that technology, which has really opened up the space frontier, once famously referred to as "the final frontier," allowing people of all nations and many different professions to journey to the Moon and to Mars. Indeed, it is these attributes that have brought your correspondent to the golden plains of Chryse Planitia to celebrate the centenary of the Space Age.

We tread a fine line in our space endeavors, between discovery of the space environment in which we live and its degradation. It would have been easy to fall on one side of the line or the other. That we did not is a testament to the men and women who ensured our safe and sustainable passage through the final frontier. As we plan our further exploration and development of the solar system, we would do well to remember their conviction and their courage.

So, what can we expect in the next 100 years? Tune in next week for INN's exclusive report from the Deimos Transfer Node ....

Chapter 10

The CASE FOR PROTECTION OF THE SPACE ENVIRONMENT

I. RECOGNITION

It has often been said that mankind's desire for exploration is one of the aspects that sets us apart from the other creatures with which we share the Earth. A natural extension of this is the desire to explore the space environment—a philosophy expressed concisely in the well-worn phrase "space . . . the final frontier." From this point of view, it is important for the long-term psychological well-being of the human race that we maintain the ability to explore space.

Moreover, as the development of industry, commerce, and tourism here on Earth has shown, our desire and ability to explore leads inexorably to a desire to develop and exploit. This strongly suggests, therefore, that exploration of the space environment will lead to its development and exploitation, and there is evidence that this is the case. Indeed, as the space environment becomes increasingly accessible, it will become a natural extension of our current business and domestic environment.[1]

As earlier chapters have shown, this ability to explore and develop brings with it an ability to pollute, degrade, and even destroy aspects of the space environment. This is, arguably, a natural extension of an ability we have developed and practiced here on Earth.

It is these abilities, and their long-term effects, that compel us to consider protection of the space environment in much the same way as the degradation of the terrestrial environmental has compelled others to call for *its* protection. The task is a difficult one in either realm—terrestrial or extraterrestrial—because those who seek to develop and exploit the environment in question are usually those with the most to lose, particularly if protection means denial of their business income. But if we agree that the space environment is worthwhile and important as an asset or resource, and therefore has a value, we automatically raise the question of its protection.

This is, by its nature, an emotive subject, which many will tend to shun because of its difficulty and the chance that it could set different parts of the space community against each other. This would obviously be counterproductive. The only solution is to take an entirely pragmatic stance. As on Earth, a key part of the solution to protection of the space environment lies in convincing the "polluters" and "degraders" that it is in their long-term business interests to protect the environment they are developing and exploiting—in effect, to protect its asset value.

No one is suggesting that protection of the space environment is easy, but then neither is space exploration—and mankind has managed to do that quite successfully. However, given the evidence reviewed earlier, it is necessary to make an effort now to make the first small steps towards a goal that would, arguably, constitute another "giant leap for mankind": the protection of the space environment for future generations of space explorers and developers.

SPACE AS AN ENVIRONMENT

The analogy with terrestrial environmentalism is convenient, and difficult to resist, given our collective understanding of "green issues." After all, we cannot build roads today without considering the fate of the natterjack toads and rare mosses, and we are encouraged to collect paper, cans, and printer cartridges for recycling, rather than add to landfill sites.

Certainly, in comparison with the other planets of the solar system, the Earth is a unique and special place and deserves protection, but the others are unique in their own right and also deserve our attention and respect. Surely, now that terrestrial environmentalism is inherent in everything we do at home, we cannot adopt a different philosophy for the other planetary bodies—not if we wish to see ourselves as a mature race, worthy of planetary exploration. We must learn from the terrestrial analog.

Of course, there is a danger of taking this convenient comparison too far, too soon. Access to the space environment—for exploration *or* development—is not simply a matter of strapping on a pair of hiking boots, packing a few supplies in a rucksack, and heading for the hills. At the moment, it requires the financial backing and human resources of national governments to transport a handful of people just a few hundred kilometres into low Earth orbit. It is therefore, politically speaking, in a very precarious position. If national governments decide to close the gateway to space, as a result of uninformed and overzealous regulation, the engine of what some call "astroenvironmentalism" will have backfired.

So before we embark on the process of protecting the space environment, we have to be clear on the ramifications of a protection regime, which depends on the extent to which we recognize space as an environment (Fig. 107).

As shown in Chapter 2, the space environment means different things to different people, from exploitable resource to object of natural beauty. Here on Earth, too, the word "environment" has come to have a number of different, though related, meanings depending on an given environmentalist's specific

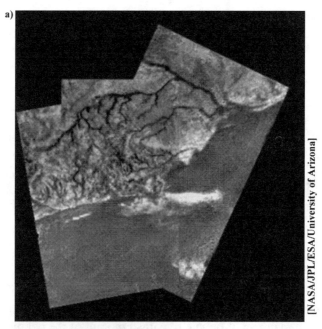

Fig. 107   Saturn's moon Titan is far from Earth: to what extent do we recognize it as "an environment"? a) dendritic drainage channels imaged during the descent of ESA's Huygens spacecraft in January 2005; b) raw image of ice blocks from the Descent Imager/Spectral Radiometer camera on Huygens following touchdown.

concerns. In fact, the terrestrial environmental movement has existed long enough to develop what one author has called "irreconcilable factions," that is, the *conservationists*, primarily concerned with the proper use of nature, and the *preservationists*, primarily concerned with the protection of nature from use.[2] Certainly, if the space environment is ever to be "used," we must recognize the semantic difference and choose conservation, or its synonym "protection," over preservation.

However, to those who do not own, operate, build, or use space systems, the space environment might have little or no meaning beyond what they see in science-fiction films, and they might care little or nothing about what happens in it, or to it. Indeed, there are many who do not realize that their telephone or Internet calls are being routed via satellite and some who find it difficult to relate a satellite weather picture to a piece of hardware in orbit around the planet, but that is inevitable. Their relative ignorance of the space environment may exclude them from any sensible discussion, but it does not mean that they will remain untouched by what happens in space (as millions of pager users across the United States discovered in 1998 when the Galaxy IV satellite succumbed to a hardware failure). In a similar way, a majority of people remains ignorant of the detailed machinations of government, but they still get taxed!

Is the ethic of environmental protection transferable to space, or is the space environment so large and so resilient to mankind's intervention that it has no need of protection? Given the facts presented in Chapters 3 and 4, the need for some degree of protection should be clear. However, many will question the relevance of the discussion, either because the exploration and development of space is of little interest to the great majority of the world's population, or because they see space as an infinite void, without form and without value. The key intention of this book is to illustrate the relevance and importance of the space environment and explain why it should be protected.

Why now? Because later might be too late. We need only point to the delay in protecting some of the Earth's most fragile ecosystems and endangered life-forms to show that much of what the terrestrial environmental movement achieved in bringing its concerns to general attention was too little, too late. This is not intended to suggest that terrestrial environmentalism has been ineffective; it is simply a recognition of human nature.

Nevertheless, in an era of growing global consciousness, in which protection of the Earth's environment is a key concern, should we not extend that concern to outer space? After all, on a universal scale the solar system is our own backyard. As a new phase of scientific and commercial development begins, with the expected Moon and Mars bases of the 21st century, should we not begin to develop guidelines for its protection, or at least an awareness of the importance of keeping it tidy?

Will the "out-of-sight, out-of-mind" rule hold sway as it has on Earth? And what about Mars? No one knows whether there is any form of simple life on Mars, perhaps hibernating in the permafrost or in the frozen water of the polar caps.[3,4] Moreover, the definition of life is continually expanding its remit, as shown by the discovery of various types of "extremophile" in the Earth's deep oceans, living in conditions formerly thought not to be conducive to life. So if life is found on Mars or Europa, it is as likely to be quite *unlike* anything we have ever seen as it is to be familiar.

If the green movement thinks it has its work cut out here on Earth, let it wait until we venture out toward the stars![5]

PROTECTION FOR THE ENVIRONMENT'S SAKE

Debris-mitigation measures, such as spacecraft deorbiting, propellant tank passivation, and cleaner separation devices, have not been developed to protect the space environment for its own sake; they are pragmatic and self-serving measures that help to protect the orbital environment for future use, either for scientific or financial gain. Protecting the space environment to maintain its qualities for commercial exploitation is surely the first line of attack in any pragmatic protection program, because it addresses the primary concern of the commercial developer: continuity of income. However, it is also worth considering protection for its own sake (Fig. 108).

Too often, we hear of irreversible environmental degradation here on Earth. As we expand our influence into the solar system, we have the potential to carry our unfortunate penchant for environmental degradation with us,

[ESA/DLR/FU Berlin (G. Neukum)]

Fig. 108 Is the Martian environment worth protecting for its own sake? (Perspective view of ice and dust layers at the Martian north pole from Mars Express).

as shown by historical evidence. Rarely has a voyage or a mission of exploration been undertaken without some detrimental impact on the environment being explored. On Earth, the exploration of the continental land masses has led, among other things, to international air travel and global tourism, which, in addition to their many advantages, make contributions to atmospheric pollution and change the original character of landscapes and indigenous populations.

Whether the advantages of exploration outweigh the disadvantages is not the issue here; it is simply a statement of fact that exploration has an impact on the environment. Of course, when exploration becomes adventure tourism, the risk of pollution becomes greater, whether on Earth or in space (Fig. 109).

Polar explorer Ranulph Fiennes would probably agree. "The worst threat to Antarctica's wildlife is posed by the ever-increasing number of cruise ship tourists," says Fiennes, explaining that a ship carrying 200,000 gallons of oil ran aground in 1989, spilling more than half immediately and the rest when the hull broke up. However, he has also experienced an alternative scenario in which conservationists succeeded in protecting Arizona's Grand Canyon from some of its most adventurous tourists. On joining a 10-day rafting tour down the Colorado River, he found it "heartening and impressive" that, despite the 15,000 annual tourists, he spotted not "a single trace of previous human visitors."[6]

Although difficult to justify to the Earth-bound majority, the space environment deserves at least as much consideration as the terrestrial environment, which in recent years has benefited from a growing public and political understanding of its fragility. In fact, if anything, many aspects of the space environment are more fragile than the Earth's because they lack the Earth's capability for self-repair. So, as space exploration evolves into space development—from commercial satellites in geostationary orbit to mineral extraction plants on Mars—the need for an understanding and appreciation of these alien environments will grow.

NEED FOR BALANCE

With the desire for exploration of the space environment on one side and the need for protection on the other, it is necessary to strike a balance.

And here lies the point of protection of the space environment: it is not about banning the use of that environment; it is concerned with ensuring that the actions of one group of users do not make it inaccessible to or unusable by others. By way of terrestrial analogy, this is like saying that protecting a marine environment would allow fishing to continue, protecting a highway network would allow commerce to continue, and protecting a football pitch would allow competition to continue. It is a purely pragmatic stance.

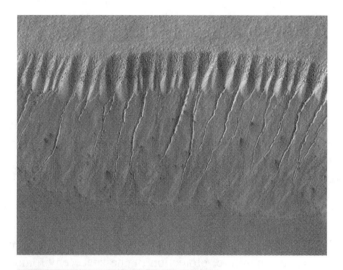

Fig. 109   Could the canyons and craters of Mars one day become tourist destinations?

This is not to say that aspects of that environment are not worthy of protection for their own sake, in the same way that one protects national parks, wilderness areas, or sites of special scientific interest here on Earth. But terrestrial experience shows that compromise is necessary: a nation cannot function solely as an environmental museum, and neither can space.

This means that exploration and development of the planetary bodies should be conducted with a view to the rights of future explorers and developers. The alternative is an anarchic approach that allows those fortunate to be first on the scene to pillage and despoil a planetary environment.

So, in the same way that a good civil engineer will consider the environmental impact of a terrestrial development, space engineers should consider the effects of their creations on the space environment. Most terrestrial development plans incorporate an environmental impact statement, so why not have one for space developments? Likewise, here on Earth, we are now familiar with the terms environmental awareness and sustainability, terms that are readily transferable to the space environment.

Some damage to the space environment is, of course, inevitable, but with forethought and understanding damage can be limited. Within a responsible and mature society, therefore, protection of the space environment should be a key consideration for any space exploration or development program. The challenge is the conception of a sustainable and environmentally aware model for space exploration and development.

Unfortunately, lack of funding and political will has placed an effective moratorium on manned planetary exploration since the days of Apollo, and no one has been able to venture beyond low Earth orbit since November 1972, when Apollo 17 returned from the Moon. This more-than-three-decade hiatus in manned planetary exploration, which could evolve into a five-decade hiatus before astronauts return to the Moon, has served to take the pressure off planners and policy makers. The only recent indication of a change in the status quo regarding manned lunar exploration came in January 2004, when U.S. President Bush announced his intention that NASA should return astronauts to the Moon by 2020 and then move on to Mars[7] (Fig. 110).

However, as long as manned lunar or Mars exploration is little more than a concept or an intention, many would say that it is premature to begin developing constraints on activities and rules of conduct for explorers, let alone developers. But a failure to place a value on planetary environments and consider their protection now could have irreversible effects.

This is why some have argued, by analogy with Principle 15 of the Rio Declaration, that we take a "precautionary approach." The Declaration states the following:

"In order to protect the environment, the precautionary approach shall be widely applied by States according to their capabilities. Where there are threats of serious or irreversible damage, lack of full scientific certainty shall not be used as a reason for postponing cost-effective measures to prevent environmental degradation."[8]

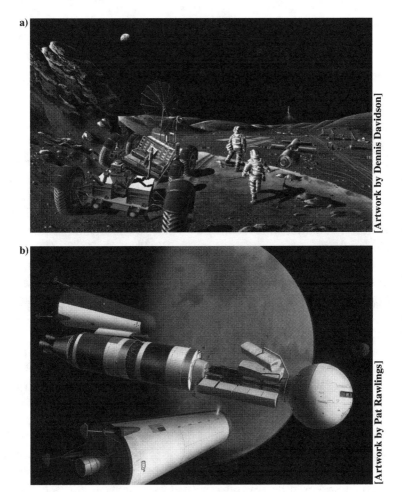

Fig. 110   a) NASA plans a return to the Moon...; b) ...and a Manned expedition to Mars.

Although the phrase "according to their capabilities" could be construed as a get-out clause, this is exactly the sort of universally agreed declaration we need for the space environment!

II. NEED FOR ACTION

EARTH ORBITS

It would be wrong to suggest that nothing has been done to protect the space environment. For the most popular Earth orbits, the case has already been made and broadly agreed upon for debris-mitigation measures. As a result, we expect to continue using low-, medium-, and high-altitude orbits for at least the next half-century without experiencing significant problems. However, the future safety and usability of Earth orbital space will depend on the continuation and improvement of the mitigation measures because a single cascade event could make a given orbit unusable, perhaps for thousands of years.

Despite the positive moves towards mitigation, and a good deal of talking about how to improve the situation, the story of orbital debris to date is mostly bad news. Moreover, current mitigation measures for the higher orbits, mainly graveyard boosts for geostationary satellites, simply delay the eventual solution. At a realistic launch rate of 20–25 geostationary satellites per year, between 1000 and 1250 spacecraft would be added to the GEO population over a 50-year period. Even assuming longer operational lifetimes for the newer satellites, this means that some 10–20 defunct satellites will be added to the graveyard each year (i.e., 500–1000 in 50 years).

The fact that these satellites are, almost by definition, abandoned and uncontrolled means that one day a collision will occur in a graveyard orbit, producing debris in orbits that will intersect the geostationary ring and threaten operational satellites. The probability of such a collision can only increase until the graveyard population is reduced.

At present, the technology to remove defunct satellites from high-altitude orbits does not exist, nor are there any funded programs to develop this technology. A significant part of any policy advocating protection of the space environment would, by necessity, include consideration of this matter.

Moreover, if new orbital applications become popular with their intended clientele, the situation could become much worse. For example, the U.S. company Celestis has become famous for its "space burial" services, which include its *Earthview* service designed to place cremated remains in orbit. This foresees urns several centimeters long "circling the Earth on average for more than 50 years."[9] Who is liable for the damage if an urn collides with another space object? The legal situation is unclear. More to the point, even if liability can be proved and admitted, what possible reparation can be made for an unusable orbit?

It has been suggested, perhaps with tongue in cheek, that it might be possible to place the burial containers in "special graveyard orbits."[9] Unfortunately, this is even worse than placing defunct geostationary satellite in graveyard orbits because the object count would eventually be much higher.

Another, oft-quoted example of the commercial use of LEO is advertising. Commercial advertisements have already been filmed in space; how long will it be before the first physical advert, visible from Earth, is placed in orbit? Although at first this sounds like science fiction—because one imagines a flickering neon sign—the advert does not have to be readable from Earth. It could simply be a flashing beacon, or a passive solar reflector like the Echo balloons of the 1960s, used as a tie-in with terrestrial

Fig. 111   The Echo space communications balloons of the early 1960s were visible from Earth—could future, illuminated advertising balloons distract astronomers?

advertising campaigns (Fig. 111). Even the diminutive Sputnik 1, an effective advert for Soviet technological prowess if ever there was one, was observable from Earth.

Advertisement hoardings, lights, and signage on Earth provoke environmental concern and so will their counterparts in space. The concept of an advert capable of being seen by a large proportion of the Earth's population might be an adman's dream, but it is an astronomer's nightmare. Likewise, considering the increasing concern about space debris, "orbiting billboards" might gain little favor with astronauts!

However, some believe that temptation and financial return might be too much to ignore: "Imagine the publicity potential," writes one observer.

> Advertisers will no longer need to worry about how many people are watching a particular television programme or tuning into an individual radio station, they will be able to broadcast to an entire nation.... In this high-tech version of delivering junk mail, the only respite we will get from the advertisers will be on a cloudy night![10]

The only consolation, in the face of this nightmare vision of inescapable intrusion, is that these days few in the developed world look up at the night sky anymore, so the application would probably be short-lived.

Another space application that is often discussed is the solar-power satellite, a spacecraft designed to collect energy from the Sun, convert it to microwave energy, and beam it down to large collectors on Earth. Ultimately, if it proves feasible and cost effective, such a system could provide enormous benefit to Earth in providing a clean, unlimited source of power while reducing our reliance on fossil fuels. The downside, of course, is that construction of such large spacecraft will be difficult, expensive, and time consuming, as illustrated by the ISS program, and can lead to an increase in orbital debris (as a result of accidents during construction perhaps). Once operational, the satellites would present a large target for existing space debris and, if they became uncontrollable, could become a debris problem in their own right. True, this is a worst-case scenario, but it should be considered in parallel with the advantages that, in a commercial environment, will undoubtedly be oversold. As suggested earlier, all such program proposals would need to include an environmental impact statement.

Beyond the commercial world, a key concern is that LEO will become increasingly militarized, particularly should one or more misguided governments decide to station a weapon system there. Possibly the most famous example of a military space program was U.S. President Reagan's Strategic Defense Initiative (SDI) of the 1980s, a proposal for a defensive shield against incoming-missile attack. The fact that SDI was more commonly known as the "Star Wars" program illustrates concerns that defense could so easily include offence, a view exacerbated by the necessarily secretive nature of military developments.

Although SDI is long dead, the sons and daughters of SDI are still with us, as shown by America's decision to unilaterally opt out of the 1972 Anti-Ballistic Missile (ABM) treaty with Russia. It did so on 14 June 2002. The next day, with the treaty consigned to history, the United States began constructing missile silos at Fort Greely, Alaska—a less than encouraging development. The facility is part of the U.S. Missile Defense Agency (MDA) Ground-Based Midcourse Defense (GMD) system, designed to provide a protective umbrella against limited ballistic missile attacks on continental North America. The first of a planned total of 16 missiles was installed on 22 July 2004.

Although official military sources argue that this and other developments are entirely defensive in nature, they continually provoke debate regarding what constitutes a space weapon.[11,12] The development and deployment, in January 2003, of the XSS-10, a small satellite designed to rendezvous with and image another object in orbit, is a case in point: if an object can be approached with some precision, it can also be hit, disabled, or even destroyed in a close-proximity explosion. The unintentional collision between NASA's Demonstration of Autonomous Rendezvous Technology (DART) spacecraft and a retired U.S. military satellite, in 2005, proved the point (see Fig. 40).

If we are brutally honest about the developments in the near future, we have to admit that the orbital-debris situation will undoubtedly get worse before it gets better. Like the proverbial oil tanker, it will need time, distance, and significant effort to turn it around. Vested interests in the space industry, agencies, and governments are likely to resist change for as long as they can because change is always difficult. But change there will have to be. The Space Age is half a century old, and, under a "business as usual" policy, it could be over for some Earth orbits within another half-century.

Moreover, if we cannot control debris in Earth orbit, which on the scale of the solar system is on our doorstep, how will we ever be able to limit debris in orbit around the Moon or Mars?

PLANETARY ORBITS

As with all aspects of space development, it is important to realize that the situation does not remain static for long. Simply because most of our spacecraft are confined to Earth orbit today does not mean that this will always be the case.

Logically, recent experience with vehicle explosions and collisions in LEO should argue for a concerted and coordinated policy of debris mitigation throughout the rest of the solar system. This is especially important if mankind intends to return to explore and develop the Moon and Mars, because a point will be reached when the volume of traffic in orbit around these bodies becomes sufficiently significant for the production of orbital debris to be considered.

It is difficult to extrapolate the potential for further degradation of planetary orbits and surfaces into the second half-century of the Space Age (2007–2057) because no one knows with any certainty what missions will be launched in that period. However, it is possible to predict the *likely* consequences of *currently expected* progress in space exploration and development.

For example, proposals have already been made to place a network of communications satellites in orbit around Mars. In the first instance, they would be used to transfer data from science stations on the surface, or flying in the Martian atmosphere, back to Earth; later they could serve as relay stations supporting a manned mission. There is no reason why areostationary orbit—the Mars equivalent of GEO—should not become an important resource in the future, eventually requiring the same protection as GEO (Fig. 112).

Perhaps before that, standardized orbital tracks around the Moon can be developed for communications, lunar imaging, and manned applications. It will then be important—not just desirable—to protect that orbital resource. NASA's 1960s policy deorbiting the Lunar Orbiters to avoid potential conflict with the Apollo spacecraft shows that a precedent for orbital protection has already been set. Unfortunately, the same pragmatism

Fig. 112   Mars Reconnaissance Orbiter—the beginning of an orbital infrastructure around Mars?

applied to debris mitigation in Earth orbit has not been extended to the planetary bodies, and there are no similar debris-mitigation measures for spacecraft that orbit those bodies.

The increase in the number of Mars probes in the early 21st century, coupled with U.S. plans to return astronauts to the Moon, shows that time is of the essence if the orbits of the planetary bodies are to be protected for future use. It is time to be pragmatic about the future use of space and realize that the solution to the orbital-debris problem—as a first priority— involves effort, cooperation, and expense.

PLANETARY SURFACES

And what about the debris deposited on the surfaces of the planetary bodies—the "forgotten space debris" such as crashed lunar spacecraft— and the effect it can have on future space exploration and development? Although the catalog of a few dozen impacts can be viewed as insignificant on a planetary scale, that catalog will continue to expand in the absence of a policy against impacts.

The only reason there have not been more intentional impacts is the complete absence of lunar missions in the 1980s and the fact that there were only two in the 1990s. (There were no lunar missions conducted between Luna

24's landing in August 1976 and the launch of Japan's Hiten/Hagoromo spacecraft in January 1990; Clementine and Lunar Prospector were launched in 1994 and 1998, respectively.)

Likewise, the only reason the other bodies of the solar system are not littered with debris from manned and unmanned spacecraft is the dearth of missions. With the single exception of targeting the Galileo orbiter into Jupiter's atmosphere to avoid a possible contaminating impact with Europa, there is little indication that these bodies are afforded any protection against spacecraft impact in the pursuit of scientific exploration (Fig. 113). What hope then when they become the target of commercial exploitation?

So far, the environmental damage caused to planetary surfaces has received very little publicity, at least partly because the damage is caused by scientific spacecraft, which tend to be seen as "benevolent" as opposed to "commercial" or "exploitative." However, the spacecraft debris that already litters the lunar surface, discussed at length in Chapter 4, provides evidence against this rather rose-tinted view.[13]

As with Earth orbit, entrepreneurs are already keen to highlight the commercial possibilities. For example, the Celestis company's *Lunar Service* is intended to place cremated remains in lunar orbit or on the lunar surface. In effect, it began in 1998 with the launching of NASA's Lunar Prospector, which included a Celestis flight capsule containing a symbolic portion of

[NASA]

Fig. 113   Mars Pathfinder enclosed in landing bags, just released from its heat shield and parachute system—the debris that accompanies a science mission.

254 SPACE: THE FRAGILE FRONTIER

the cremated remains of lunar geologist Eugene Shoemaker.[9] Given the mission's potential for introducing biological material to the lunar south pole, what guarantee is there that the Moon's environment will not be adversely affected by future missions?

Indeed, a third Celestis service, named *Voyager*, which aims to launch cremated remains into deep space, begs further questions. Although, according to the company, its passengers should "travel harmlessly and eternally through the vastness of space," the indefinite duration of the mission results in a legal problem under the Liability Convention because it implies an "indefinite duty of jurisdiction and control of the state of registry."[9] The fate of a spacecraft released into the vastness of the solar system might seem somewhat academic today, but the fact that a Saturn V rocket stage could suddenly reappear after 30 years to potentially impact the Moon sets a less academic tone[14] (see Chapter 4).

It seems even more likely that the planetary surfaces could provide advertising opportunities. As a public relations stunt, the Mars lander Beagle 2 included a painting by Damien Hurst as part of its payload, ostensibly to calibrate the spacecraft's color camera (Fig. 114). It is not difficult to imagine future lander cameras being calibrated using McDonalds or Burger King logos, or missions being sponsored by companies in need of a new advertising angle: the Mars Confectionery Orbiter and the Sol Beer Solar Probe, perhaps.[10] Although such initiatives might only pollute the radio spectrum, it falls to those interested in protecting the space environment to recognize the possibilities.

More controversial still would be plans to use the Moon as a nuclear waste repository, as suggested by those against America's plans to use Yucca Mountain, Nevada, as a burial site for 77,000 of tons of radioactive material.[15] According to Sherwin Gormly, an environmental engineer from Nevada, redundant ICBMs tipped with their nuclear payload could be fired, using a proposed suborbital space plane, towards a specified crater on the Moon, where they would bury themselves under the lunar regolith. The argument goes that although the impact area would be highly contaminated, the problem of waste migration would be eliminated because the Moon has no hydrosphere. However, the waste-filled casks would probably not bury themselves conveniently under the surface but, instead, form sprays of ejecta that would distribute the material across the surface, and possibly into lunar orbit. Apart from that, of course, the reliability of current launch vehicles is much too low to risk carrying such a payload; if it was otherwise, the far more preferable "crazy idea" of launching nuclear waste into the Sun might have been developed by now.

Interestingly, Gormly admits that the reaction against the Nevada proposal is a typical "not in my backyard" reaction, adding that "NIMBY politics don't apply to the lunar surface at this time." But this is exactly

[ESA/Denman productions]

Fig. 114   Beagle 2, the ill-fated Mars Express lander with which contact was lost prior to landing—the forerunner of advertising-sponsored missions?

the point: the Moon *is* in our backyard, and we should apply at least a degree of "NIMBY politics" to it. The difficulty in convincing most people that this is so is twofold: they cannot see the lunar surface as an accessible backyard because no one has been there in more than 30 years, and they cannot *see* the lunar surface in detail because it is so far away. This makes a policy of "out of sight, out of mind" considerably easier to establish than it is for the Earth's surface environment.

It is important that this myopia regarding the planetary bodies should not be allowed to remove responsibility from those involved in exploring and later developing those bodies. Moreover, it would be advisable to consider this now, as opposed to waiting until there is a need for remedial action, as there is already in LEO. The old adage "prevention is better than cure" is particularly apposite.

The problem, of course, is obtaining a consensus. Research has shown that some people consider planetary bodies other than the Earth to be barren, unwelcoming worlds that are either "ripe for plunder" in terms of material wealth, or need to be "civilized," perhaps by terraforming.

This opinion is analogous to the American public's view of nature in the nation's early history, when, according to author Howard McCurdy, wilderness areas were considered "savage, uncontrollable, and evil."[16] It was

Fig. 115   Artists' impressions help to romanticise space, but reality has an undeniable beauty of its own: a) geologist-astronauts explore Mars; b) Saturn and its rings from Cassini.

not until landscape paintings, which romanticized those areas as places of great natural beauty, were exhibited in the early 19th century that the concept of conservation became realistic. Later that century, paintings of the Rocky Mountains and Yosemite Valley helped build support for the national park movement, and Congress was even moved to appropriate $20,000 for similar examples to hang in the U.S. Capitol, according to McCurdy.

However, although many people have an inherent appreciation of the need for planetary protection, many more remain to be convinced (Fig. 115).

Currently, as far as the space environment is concerned, it is not the general public that needs convincing; it is the space professionals and organizations that procure, design, and finance space hardware that must be made aware of the need to protect planetary environments. In the future, when the Apollo landing sites become tourist attractions—perhaps even international parks—the visiting public will thank the space professionals for their foresight.

However, it might be too early in the game to begin designating protected areas now, as philosopher Holmes Rolston points out: "Speculating over what places, planets, moons should be designated as nature preserves would be more foolish than for Columbus to have worried over what areas of the New World should be set aside as national parks and wilderness." Nevertheless, he concedes, "in retrospect, our forefathers would have left us a better New World had they been concerned sooner about preserving what they found there...."[17]

As with the Earth-bound environmental movement, there is a need to raise awareness among both space professionals and the general public. However, terrestrial experience of "ecowarriors" and "greens" of various shades argues for caution because overzealous and inaccurate claims broadcast in emotive language can lead to alienation rather than understanding.

The intention must be to provide education without alienation, perhaps taking the line of the 19th century wilderness painters quoted by McCurdy. It will not be easy, however. Whereas photographs of the Earth from space, particularly those resulting from the Apollo missions, had an important affect on society, arguably even kick-starting the environmental movement, images of the more distant planetary bodies are unlikely to provide such a fundamental boost to the planetary environment movement. Whereas everyone who lives on the Earth has a vested interest in protecting their home planet's environment, no such interest exists for the Moon and Mars, possibly because there is no concept of ownership in the emotional sense (Fig. 116).

The emergence of constructive proprietorial feelings might have to await the development of space tourism, at which point a vested interest in preserving both planetary surfaces and their historic sites might develop. Of course the development of tourism facilities is not without its environmental concerns, and it might be that legislation is the only recourse in protecting the space environment.[13]

The aim of those interested in protecting planetary environments is that the issue of planetary debris should be recognized among space professionals in the same way that orbital debris is today. Once that aim is achieved, awareness of the problems, and their solutions, can be extended to future users, owners, and developers of the space environment.

This raises the vexed question of whether parts of the space environment can be owned at all (see Chapter 5). Certainly, so far, no part of it is legally owned by any person or state. Even the geostationary orbital positions used exclusively by their current occupants are not owned by them; they are simply "on loan" from the international community and coordinated by the International Telecommunication Union (ITU). However, if Earth history is anything to go by, it seems likely that ownership of planetary

a)

[NASA/JPL/Space Science Institute]

b)

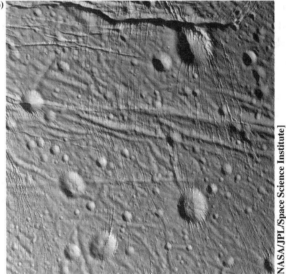

[NASA/JPL/Space Science Institute]

Fig. 116   Saturn's moons may be interesting and beautiful
in their own way, but do Earth-dwellers feel "ownership"
or responsibility towards them? a) Cassini's view of
Phoebe shows the effect of impacts on a small body;
b) Enceladus is an icy moon scarred by tectonic forces. The
10 km-wide crater at upper right has prominent north–
south fracturing along its north-eastern slope.

real estate will be acquired by laying claim. It will be acquired by actually landing there and claiming—perhaps through some legal registration process—the right to develop the estate.

Philosophers and ethicists have been considering such matters since the early years of the Space Age, generally to little practical effect. "The question," asks one, "is whether this astronomical world can *belong to us*; there is no question how we *belong to it*, and no question whether it *belongs to itself* [his italics]."[18] The answer is still far from clear.

Eventually we must decide whether the planetary environments should be declared as international parks in which all commercial and industrial development is prohibited, or whether they constitute a body of resources to plunder. The most likely outcome will be some combination of the two. Until then, perhaps we should follow the lead of the U.S. National Parks with their slogan: "take only photographs, leave only footprints."[19]

III. THE WAY FORWARD

So where do we go from here? How do we protect and preserve the fragile frontier? And how do we ensure a balance between overbearing protection and unfettered regulation?

It is important to be clear about the proposals here. They are not designed to protect the space environment by cordoning off the Earth and condemning the human race to end its days as a single planet species. To do so would ignore our natural desires to discover and explore the environment around us. It would also restrict our needs and abilities to better ourselves through technological progress and the development of our natural and built environment. Placing an embargo on space exploration and development would be like confining the inquisitive caveman to the valley of his birth.

Nevertheless, the activities of mankind throughout history have proved the need for guidelines, regulations, and legislation that instruct and govern our Earth-bound pursuits. There is no reason to believe that we can do without them in space.

Indeed, the fact that a new kind of space activity can begin without significant legal or political preparations for protection is shown by the visit to the ISS of the first space tourist, Dennis Tito, in April 2001.[20] The visit was, effectively, a private deal between a government and a wealthy individual, brokered by a commercial third party. There were plenty of people in other government bodies with interests in the station who were against the flight of a civilian to the ISS, not least those in ESA who thought their trained astronauts should have priority, but Tito's flight went ahead regardless. That is not to say it was a bad thing, simply that it was organized outside the recognized framework for space activity (limited though that framework might be).

The second tourist flight, of Mark Shuttleworth in 2002, showed how quickly space tourism could develop once the infrastructure is in place. The main reason that further "spare seats" on Soyuz ferry flights to the ISS were not immediately taken by fare-paying passengers was the upheaval created by the Columbia accident. With transport options limited to the Soyuz and the station crew reduced to two, having tourists floating around was considered a risk too far.

Once again, extending consideration to the Moon and assuming the current intention to return astronauts to the lunar surface bears fruit, we could see a Moon base of some description in place by perhaps 2020. How long would it be before seats on ferry flights to "Moon Base Alpha" were offered to billionaires in search of the ultimate vacation? Though it seems like science fiction today, so too would Tito's flight in the days of Skylab.

Fact has a habit of overtaking fiction in this business, and we need to be ready to cope with the effects, both beneficial and otherwise, and the only way to be ready is to prepare!

Many if not most of those concerned with the protection of the space environment are themselves children of the Space Age. They grew up with thoughts of going into space and watched with wide-eyed fascination as white-suited figures gamboled about on the lunar surface. Following years of investment in formal education and work experience, they have expert knowledge in space science, technology, policy, or some other, related field. Today, their livelihoods, and those of their families, depend on the continuation of space applications and development, and not unlike the would-be developers, they have a vested interest. They are not about to barricade the space frontier and turn their backs on everything that has driven their careers and their lives.

Rick Steiner, of the University of Alaska, summarizes this with some eloquence: "we *should* expand into our universe, but *not* as greedy industrialists, empire builders, or with militant intention. Rather we should expand into the universe *only* with deep respect, scientific curiosity, and humility—two fundamentally different approaches."[21]

Likewise, lawyers Patricia Sterns and Les Tennen state: "We are, after all, merely guardians of the universe for future generations, and we have, therefore, an absolute obligation to utilize the heavens wisely."[22]

PROMULGATING IDEAS

If those with an interest in protecting the space environment wish their ideas to be considered by the wider space community, while dispelling any misunderstandings of those ideas, they must work to spread the word.

Evangelising is never easy. Like brand recognition in advertising, it is a continuing background task, the gradual insinuation of an idea that something is right or good. There might be a few instantaneous conversions along the way, but those with a calling for protection of the space environment must be in it for the long haul.

At an academic level, the first small steps have already been taken. Individual space professionals have published articles and papers drawing attention to some of the issues, and a number of special conference sessions[23] have made a more formal attempt to present the issues to the international space community.

Although these attempts are laudable, they represent only a fraction of the effort required, firstly, to convince the space community of the need to consider protection and, secondly, to formulate a policy. So the onus is on those with particular expertise to share to continue talking to their peers in industry, the space agencies, and other bodies, and promulgating the issues at conferences and in publications.

If the aims of the space protection lobby are to be realized, the discussion of the subject will require a good deal more collective knowledge, understanding, and maturity than has been evident in similar discussions regarding the Earth's environment. At present, that knowledge resides largely within a small subset of the professional space community, so there is a need for promulgation both within and beyond that community.

This is no easy task because there are many who will see this as another layer of bureaucracy and yet another restriction on progress. Some might even compare the space protection lobby with the Luddite machine breakers of the early 19th century. This would, however, be controversial and counterproductive because the space protection lobby is not against the industrial development of space; it simply wishes to ensure that development takes place in a sustainable and environmentally friendly manner (Fig. 117).

Nevertheless, one of the many challenges in communicating information and views on space protection to a wider audience will be that of demonstrating that the lobbyists are not latter-day Luddites.

CONSULTATIVE BODY

Small steps aside, significant progress in any subject is made by giant leaps, and the giant leap in protection of the space environment would be the formulation and agreement of a policy.

The first step towards that goal would be the formation of an international consultative body to consider the relevant issues and raise awareness of the subject among the growing body of space professionals and practitioners.[24] The question is, of course, under whose auspices should this body be inaugurated?

As suggested in Chapter 6, an international body including representatives from all major space-faring nations should be formed, possibly under the

[NASA]

Fig. 117 It is inevitable that lunar exploration will lead to development—the challenge is to do it in a sustainable and environmentally-friendly manner.

auspices of the United Nations. We should learn, however, from the problems encountered on a daily basis by the ITU, a specialized agency of the UN, which is saddled with an overbearing bureaucracy, insufficient resources, and limited powers of enforcement. Unfortunately, any future "International Space Protection Union" might have to function under the same restrictions. The key to success will be to get as many parties with vested interests on side as soon as possible (Fig. 118).

Thus, to represent the science community, one might seek to gain backing from major space agencies, such as NASA, ESA, and the Japanese space

Fig. 118 An international body for protection of the space environment might be modelled on the International Telecommunication Union... but it would have to learn from its mistakes.

[Mark Williamson]

agency, JAXA. To represent industry, one might seek the support of national space industry associations or, better still, the leading prime contractors themselves. However, to make the organization as democratic as possible and to allow a broad-based ownership of the ideas, issues, and eventual policies, one would have to accept representations from a much wider field. This is important because we appear to be on the cusp of a true commercialization of space, and there is no point in formulating policies to which only the industrial prime contractors will adhere; there has to be a way to include the smaller operators.

The proposed organization must therefore not only strive to enhance awareness of the issues, but also to encourage discussion of all aspects, not least to allow feedback on initial ideas.

FORMULATING POLICY

So, assuming that the formation of an international consultative body goes to plan, what should it do? The first few items on the agenda must surely include the following:

1) formation and enactment of a policy to maintain and expand the constituency of the body, specifically regarding its international nature;
2) formation and enactment of a policy to obtain funding and other support from key space-related organizations;
3) formation and enactment of a policy to ensure the promulgation of ideas among the space community and the media; and

4) consideration of a set of guidelines or code of practice as a precursor to more formal policies or legislation.

Although the last point is the *raison d'être* for the body, the preceding points are also important. This can be illustrated by considering the ramifications of omitting them:

1) If the body is not international in nature, it will be considered by those outside as partisan, and, however well meant its eventual policies, they will be ignored.

2) If the body cannot secure at least some regular funding, it will be unable to operate a secretariat or provide expenses to those members unable to fund their activities in support of the body.

3) If the body does not appoint an experienced promulgator, the ideas and issues considered by the body will remain in house, much as they are today.

As far as the set of guidelines or code of practice is concerned, this should take into account the work that has already been done in formulating space laws and ethical space policies, as described in Chapters 5 and 6, respectively. It should be clear that the consultative body is not starting with a blank canvas, as the UN's ad hoc Committee on the Peaceful Uses of Outer Space (COPUOS) was obliged to at the dawn of the Space Age. We have come a long way since then.

Hopefully, the guidelines would evolve relatively quickly into a policy which space-faring governments would be willing to adopt and a set of regulations that they would be willing to enforce.

An obvious first priority must be the orbital-debris problem. It is already well-recognized and studied and would therefore be the issue most readily resolved. If the international space community were to regulate access to the space environment, placing limitations on the technology and devices allowed in orbit, it would be well on its way to halting the increase in the debris population.

Indeed, the formation of a regulatory body with the power to approve satellite and launch-vehicle designs would be a welcome innovation. Although debris mitigation is the primary concern, such a body might also be empowered to judge the wider environmental suitability of the payloads.

Although some would condemn this as a draconian, anticompetitive measure that favors those manufacturers and operators which choose to ignore the regulations, this is not a valid argument against space debris mitigation. Indeed, the possibility that some might ignore the regulations simply highlights the need for legislative powers to enforce them. A terrestrial parallel is the drive to reduce pollution from motor vehicles by fitting engine management systems and catalytic converters; it is time to instigate analogous measures to protect the space environment.[25]

In addition to halting the *increase* in orbital debris, it would be sensible to instigate an international collaborative program to reduce the *existing* debris population. This might be done, for example, by way of a joint venture between the world's space agencies, which could issue a joint request for proposals to industry for the development of a spacecraft deorbiter. This is, after all, the only way defunct satellites will ever be cleared from the geostationary graveyard orbits, so it should be seen as worthwhile expenditure.

Of course, it is easy to suggest ways in which space agencies should spend money they do not currently have, but there is no point in highlighting a problem without suggesting a solution—especially if the solution is as obvious as this one! Moreover, the technology is well within our grasp.

A future spacecraft deorbiter would probably be designed along the lines of the previously proposed orbital transfer vehicles (OTVs) or "space tugs" and would be capable of changing its orbit to rendezvous with defunct satellites and bring them down to an altitude from which they could either be deorbited (and incinerated) directly or collected for return to Earth. By incorporating, among others, the technologies of solar sailing, ion propulsion, teleoperation, dextrous manipulation, and in-orbit refueling, the spacecraft could be justified to the agencies as a technology demonstrator. Other, far-sighted individuals might see it as the precursor to a fleet of extraterrestrial snow ploughs that would keep the space lanes open through what could otherwise be a very long winter.[25]

The potential has already been demonstrated by Japan's experimental technology satellite ETS-VII, whose chaser and target spacecraft performed docking and undocking tests in July 1998. Meanwhile, the U.S. Department of Defense has been working on a satellite servicing and refueling demonstration under its Orbital Express Program,[26] and commercial European companies have developed the ConeXpress Orbital Recovery System, designed to extend the operational life of communications satellites.

As for the orbital transfer requirement, ESA is already developing its Automated Transfer Vehicle (ATV) for the ISS, which could no doubt be adapted (Fig. 119), and NASA's Langley Research Center has studied designs for an orbital architecture that includes solar electric space tugs and crew transfer vehicles.[27] Whatever the technology, if satellites are ever to be removed from the higher-altitude orbits, it is likely to be done by a commercial enterprise, even if working under contract to a government agency, which means that orbital protection could one day be a profitable commercial business, rather than an onerous requirement.

Unsurprisingly, similar ideas have been part of published science fiction for many years. For example, a children's novel of the early 1960s called *Expedition Venus* featured "Salvage Tug Number One, a specially designed space vehicle whose task it was to gather up the debris of space."[28]

Fig. 119   Could ESA's ATV be adapted as an orbital transfer vehicle?

It might be too soon to suggest that space garbage collection could be a commercial space application, but, once again, we would do well to remember that most of today's space applications began as science fiction. Dismissing the notion that defunct satellites could have a scrap value is simply turning a blind eye to the inevitable. As on Earth, any recycling initiative will have to be subject to a costs-benefits analysis, but if benefits included factors as fundamental as "the continuation of orbital activities," initiatives could readily become funded programs.

A second priority, after orbital debris, should be what we have, for sake of brevity, called planetary debris—the remains of spacecraft on the surfaces of planetary bodies. Some have suggested that this too could one day be regarded as a valuable source of raw materials, rather than trash, but that remains to be seen because of the difficulty of locating widely scattered parts. Although the cost of clearing up the debris of crashed spacecraft and rocket stages is likely to be prohibitive for some time, a significant effort could usefully be made in ensuring that additions to this debris population are minimised.

On the basis of our experience with spacecraft debris in Earth orbit alone, it is clear that the subject of planetary spacecraft debris deserves serious consideration, possibly under two headings: debris monitoring and debris mitigation. Both subjects could be studied by the proposed international consultative body, working in concert with the commercial space industry to ensure that mitigation measures were practicable and cost effective. Of course, some of the man-made objects on the planetary bodies could be designated not as debris but as heritage or museum pieces—the future "must see" destinations for space tourists (Fig. 120).

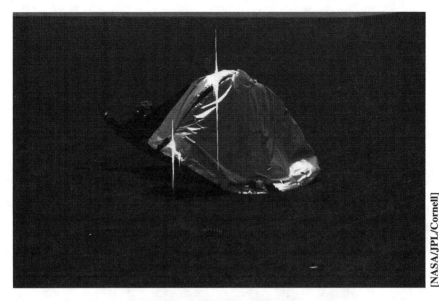

[NASA/JPL/Cornell]

Fig. 120   Is Opportunity's heat shield an engineering curiosity, a future museum artefact or discarded debris? The springs and other objects surrounding the shield represent the "unseen" environmental impact of planetary missions.

As far as protecting the planetary environments themselves is concerned, the consultative body will need to prioritize the planetary bodies and areas on those bodies with a view to their scientific importance. Astronomers Almar and Horvath[29] have made the practical suggestion of prioritizing solar-system bodies under a number of headings:

1) bodies in unusual orbits or positions (e.g., Phobos and Deimos, the only asteroid-like satellites of an Earth-like planet);
2) bodies of unusual form (e.g., some minor planets);
3) unique planetary surfaces or interiors (e.g., Io, Europa, and Valles Marineris on Mars);
4) special atmospheric and other features (e.g., Jupiter's Great Red Spot, magnetic fields, etc.);
5) historic exploration sites; and
6) places where the probability of finding fossil life-forms is nonnegligible.

Almar regrets the fact that protection of the space environment, beyond the problem of space debris, was not mentioned in the so-called Vienna Declaration on Space and Human Development, which resulted from the UNISPACE III conference.[30,31] However, he reports, a technical forum did manage to add a recommendation for "defining roles for the protection and preservation of the planetary and space environment and establishing a framework for implementation."

a)

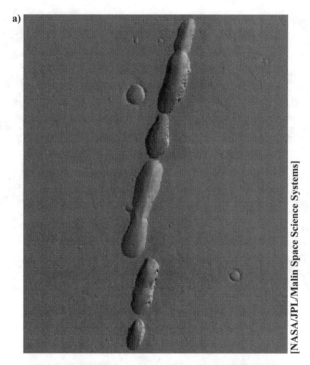

[NASA/JPL/Malin Space Science Systems]

b)

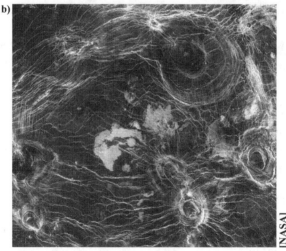

[NASA]

Fig. 121   Solar system features to prioritize for scientific importance? a) a chain of pits associated with lava flows in the Tharsis region of Mars; b) Arachnoid formations on Venus imaged by the Magellan orbiter's synthetic aperture radar (SAR) in 1990/91. So called because they resemble spiders and webs, they are thought to be volcanic features formed by the upwelling of magma and range in size from 50 to 230 km in diameter.

He concludes that COSPAR would be best placed to establish a panel to categorize and prioritize the sites of special scientific interest (Fig. 121). Finally, he says "these internationally selected scientific preserves or 'wilderness areas' should be legally protected within the frame of an international environment-protection treaty of the United Nations...open to scientific investigation but closed to exploitation of extraterrestrial resources."

Detailed suggestions concerning the content and coverage of the proposed guidelines or code of practice will undoubtedly be made in due course, as one of the next small steps towards the goal of protecting the space environment.

In formulating these guidelines, it will be important to recognize the need to strike a balance between uncontrolled exploration and development and a stifling regime of rules and regulations. It is not in the nature of mankind to open a new frontier and then close it again without fully exploring its potential. Specifically, it is not in mankind's best interests to place a moratorium on space exploration and development.

The challenge for 21st century space practitioners will be to find a way to explore, develop, and exploit the unique properties of the space environment, including the surfaces and orbital spaces of the planetary bodies, without degrading or destroying the environment that holds so much promise for the future of mankind.

POLICING THE POLICIES

Once the policies have been decided and agreed upon—no small feat in itself—the community will face the thorny issues of legislation, policing, conviction, and penalty. If an operator or developer ignores the policies and breaks the rules, will there be a mechanism to cope with this? Who will pay for pollution or degradation of the space environment, and how will this be policed? (Fig. 122).

Here on Earth, we often uphold the ethic of "polluter pays." However, in terms of environmental policy this is often too little, too late: for example, when a major oil spill occurs, it is not only the carrier who pays for the clean up; in a sense, the environment pays too. Disaster mitigation is a key policy in this context, and it is widely understood that oil tankers should have double hulls to help contain their cargo in the event of a collision. Naturally, this has a cost impact on the carrier and, ultimately, on the consumer.

Certainly, a policy of polluter pays would be too late for the space environment because, in most cases, the damage done could not be corrected. Once an orbit is clogged with spacecraft debris, or a planetary surface contaminated, it becomes extremely expensive, if not impossible, to clean it up. A debris-mitigation policy should therefore be an integral part of any pragmatic policy, if only because it would be more cost-effective for

[Willis Inspace]

Fig. 122   How can we police orbital pollution?

the space industry in the long run. The challenge, of course, is to convince the industry that this is so.

Moreover, finding money for something after the event is more difficult than drawing on a dedicated budget. In this regard, space law expert Bess Reijnen believes that an international fund should be established "for clean-up operations in outer space."[32] It could, she says, be based on mandatory contributions from the launching state of registry paid prior to launch. Naturally, this will be seen by detractors as a tax on space operations, but we should not be surprised if the maintenance of our orbital infrastructure attracts a tax in the same way that we pay to maintain our terrestrial transport infrastructure. Although initially this would increase the cost of space access, which no one wants, it would be preferable to having *no space to access* should an orbit become unusable.

Any move to change the status quo, in any walk of life, is problematic, and any initiative to protect the space environment would be no exception. Not only is the space environment "out there," divorced from everyday experience, but also "long term" in comparison to anything else we know. Although terrestrial features such as lakes, rivers, and glaciers seem permanent to human observers, they are mere blinks in the eye of creation compared with the orbital progression of the planets. Although rivers can be polluted within a fraction of a human lifetime, a body the size of a planet is so large that a single act of pollution would seemingly go unnoticed.

Fig. 123 Buzz Aldrin described the Moon in terms of "magnificent desolation." Examples of this "desolation" should be protected for future visitors.

In the same way that migrants exploited parts of the American West and moved on, exploiters of the space environment could arguably do the same. In the absence of a rigorously enforced clean-up policy, they would probably be inclined to reap the benefits and then develop another site, rather than being stuck with the reparation costs. And who would be there to stop them?

What is required for the sustainable and environmentally friendly development of space is a change of mindset. It requires a rejection of short termism and an embrace of what has been called "multigenerational altruism," in this case a long-term unselfish concern for the space environment (Lupisella, M., personal communication, 5 May 2004).

Many of the measures necessary to protect the space environment require a good deal of imagination and forward thinking, and, most importantly, a candid admission of the worst-case scenario. Paradoxically, for a community famous for its powers of imagination, the space profession has a habit of viewing the future through rose-colored spectacles.

And although the space community's record on protecting its own operational environment is poor, its reaction to recommendations on how it should do so is, at best, unsatisfactory. For example, figures show that the ITU recommendation to boost defunct communications satellites to a 300-km graveyard orbit has not been followed by many operators. As reported in Chapter 3 (see Box: The Altitude of the Graveyard Orbit), more than 60 satellites were left in or near geostationary orbit between 1996 and 2001, creating a risk of collision. Despite the success of the ITU in coordinating the use of orbital and frequency space, this shows that voluntary regulation does not always work.

The same could be said for the body of space law, as discussed in Chapter 5. As astronomer Lubos Perek points out,[33] in May 2002, the On-Line Index of Objects Launched into Outer Space, prepared by the UN Office of Outer Space Affairs, showed that "30% of objects launched in the year 2000 have not been registered with the UN, in spite of Article XX of the [Registration] Convention." Moreover, it appears that the number of payloads actually registered is decreasing as a percentage of those launched: between 1976 and 1980, for example, 99% were registered, but this had decreased to 91% for the 1991–1995 period, and was only 78% for 1996–2002.[34] These figures appear to confirm the opinion, held by some, that space is an unregulated vacuum, devoid of law, responsibility, and common sense.

Taking the stance that a formal body to authorize and control space activity is required, Bess Reijnen believes that an Outer Space Authority should be established, specifically with regard to "the orderly and safe development of the Moon."[35] She suggests that lessons learned from the Deep Seabed Authority, which was established when exploration and exploitation of the deep seabed became technically possible, and other aspects of the law of the sea should be transferable to space. Although moves to establish such an authority for outer space put forward in the 1970s and 1980s came to nothing, Reijnen suggests that proposals to protect certain lunar craters for radio astronomy and SETI applications might provide the stimulus for another attempt. "The requirements of the astronomical community," she says, "would seem best served by the creation of such an Outer Space Authority," which could also help to enforce the existing body of space law.

Although, in current usage, the term "space control" implies a control of space from a military perspective—including such aspects as surveillance, missile warning, and antisatellite weapons—it has a certain resonance in the present discussion. According to space policy expert Ray Williamson, "the concept of space control involves complex interactions in the realms of international politics and law, economics, and national security."[36] The potential for disabling or destroying an adversary's space systems, he says, might be "too much of an attraction to pass up during time of conflict ... but the unintended physical, economic, and political consequences may be great."

It matters little to the space environment whether orbital debris results from an accidental collision between two commercial satellites or the intentional detonation of an ASAT weapon. The result is the same. And what we learn to do or allow to happen now, in Earth orbit, will soon be directly transferable to the Moon—its surface and its orbital space.

In contrast to those who like to draw analogies with terrestrial laws, systems, and culture, Ray Williamson adopts an alternative viewpoint. "Outer space is truly a different kind of place, with characteristics and a political history different from the land, sea, and ice," he says. "Simplified solutions based on historical analogues ... will not suffice."

Although this may be true, outer space is being explored and developed by the same human race that has explored and developed the Earth. Experience shows that humanity, in general, requires a degree of regulation to ensure that it does the right thing and that this regulation requires enforcement. To ignore this truism is tantamount to ignoring human nature.

MORAL RESPONSIBILITY

All of the factual information concerning orbital debris and the effect of spacecraft on the planetary bodies contained within this volume has been available to members of the space profession for several years. Although they might not have assimilated this knowledge, its free availability brings with it a degree of moral responsibility that cannot be ignored by those who consider themselves professional. To do so would be unethical.

Moreover, there comes a point along the path of discovery in any subject where those who understand the issues, some of which might not be entirely palatable, must accept a degree of responsibility for the information they hold. Knowledge, once learned, can be ignored, but it cannot be unlearned. For example, individuals who learn of a colleague's misdemeanor can ignore it, but they cannot easily wipe it from their memory or their conscience. In an ethical environment, individuals must decide what to do with the information, check it if necessary, place it in context, and balance the pros and cons of sharing it, but they cannot ignore it.

Such a dilemma was faced by the American marine biologist Rachel Carson in 1957, when she received a letter from a friend concerned about the spraying of the pesticide DDT from aircraft in an attempt to control the mosquito population. Apparently, the day after the friend's woods were sprayed, there were dead songbirds in her yard. Carson's research led, in 1962, to the publication of *Silent Spring*,[37] a book widely credited with launching the modern environmental movement.

Although Carson was a respected and accomplished scientist and author, her work was vehemently opposed and criticized by vested interests, including chemical companies and the bodies and individuals they subsidized. Nevertheless, Carson was vindicated, albeit posthumously in 1969, when the U.S. Congress passed the National Environmental Policy Act and the National Cancer Institute announced its findings that DDT could produce cancer. By then, individual states had started to ban the use of DDT, and, in 1972, a federal ban was placed on the pesticide.

Rachel Carson made a difficult decision in assuming the role of whistle-blower, but few would deny today that her decision was morally correct. Depending on one's perspective, it took 10, 20, or even 30 years for the world to fully appreciate Carson's work; some would say it is still not fully appreciated. This shows that it is never too early to warn of potential dangers.

Interestingly, in the present context, the introduction to the 1965 United Kingdom edition of *Silent Spring* by Lord Shackleton contains a quote from Prince Bernhard of the Netherlands, speaking at a Wildlife Fund dinner in London: "We are dreaming of conquering space. We are already preparing the conquest of the moon. But if we are going to treat other planets as we are treating our own, we had better leave the Moon, Mars, and Venus strictly alone!"[38]

More than 40 years after these words were spoken, we are in a position to consider this recommendation with the benefit of hindsight and specific knowledge of the effects of space exploration. The lessons learned on Earth should be directly transferable to space (Figs. 124 and 125).

Unfortunately, many people who consider themselves space professionals would prefer all this talk of protection to go away, so that they can concentrate on exploring the solar system, developing technology, and making money. Sadly for them, this is not going to happen, because a continuation of the laissez faire policy will result, eventually, in a depreciation of their key asset—the space environment.

[©David A. Hardy/www.astroart.org]

Fig. 124   Space is part of our environment and the lessons learnt on Earth should be transferable to space.

Fig. 125   Will the first person to watch a Martian sunset consider Mars an exploitable asset . . . or a part of the space environment that should be protected? Only time will tell. (Sunset from NASA's Mars Exploration Rover Spirit on 19 May 2005).

Indeed, if financial gain were to be the sole driver for the desirable and inevitable expansion of mankind into the solar system and beyond, we should wait until we have gained the degree of maturity and responsibility that befits this momentous and important goal.

It is always difficult to go against convention, especially when you risk biting the hand that feeds you, but when the future of space exploration and development is at risk it must be worthwhile.

One day, under the most optimistic scenario, an acceptable balance will be struck in which the advantages of space exploration outweigh the disadvantages. This will not happen, however, under a policy of laissez faire. Although the space environment has yet to reach the point of no return, restrictions on the use of parts of that environment and the irreversible degradation of others might be closer than we think.

As space becomes accessible to a wider variety of individuals, corporations, and other bodies, the need for protection of the space environment grows. If the space environment is to remain available for the study of and use by successive generations of explorers and developers, we must make serious steps towards protection. An effort to develop a set of guidelines for the future exploration and development of space must be made now— before "the final frontier" becomes a lawless, selfish, and untamed frontier.

In another 20 years or so—when the second generation of lunar explorers is making footprints on the surface—it might be too late.

REFERENCES

[1]Williamson, M., "Exploration and Protection—A Delicate Balance," International Academy of Astronautics, IAA-01-IAA.13.1.03, Oct. 2001.

[2]Hargrove, E. C. (ed.), *Beyond Spaceship Earth: Environmental Ethics and the Solar System*, Sierra Club Books, San Francisco, 1986, p. XII.

[3]"Hydrogen-Fed Bacteria May Exist Beyond Earth," NASA Press Release 02-37 AR, (3 April, 2002).

[4]"NASA Scientists Create Amino Acids in Deep-Space-Like Environment," NASA Press Release 02-33 AR, (27 March, 2002).

[5]Williamson, M., "Space and the Environment," *Engineering Science and Education Journal (IEE)*, Vol. 1, No. 3, June 1992, pp. 145–151.

[6]Fiennes, R., *Mind over Matter: the Epic Crossing of the Antarctic Continent*, Sinclair-Stevenson, London, 1993, p. 159.

[7]Williamson, M., "Seeing Red," Analysis, *IEE Review*, Vol. 50, No. 2, Feb. 2004, pp. 22–25, URL: http//:www.nasa.gov.

[8]Bohlmann, U., "Planetary Protection in Public International Law," IAF IAC-03-IISL.1.05, Oct. 2003.

[9]Hofmann, M., "Legal Framework of Space Burials," *Earth Space Review*, Vol. 10, No. 2, 2001, pp. 64–70.

[10]Wade, D., "Directions in Space: a Possible Future for the Space Industry," Part I, *International Space Review*, Dec. 2004, pp. 6–8; Part II, Jan. 2005, pp. 8,9, URL: www.satellite-evolution.com.

[11]Hitchens, T., "When Is a Space Weapon not a Space Weapon?," Commentary, *Space News*, 19 Jan. 2004, p. A8.

[12]Samson, V., "Lift the Veil on Space Weaponization," Commentary, *Space News*, 12 April 2004, p. 13.

[13]Williamson, M., "Planetary Spacecraft Debris—The Case for Protecting the Space Environment," *Acta Astronautica*, Vol. 47, Nos. 2–9, 2000, pp. 719–729, also International Academy of Astronautics, Paper IAA-99-IAA.7.1.01, 1999.

[14]Britt, R. R., "Earth's Newest Satellite Could Hit Moon Next Year," Space.com, [cited 12 Sep. 2002].

[15]David, L., "Moon Seen as Nuclear Waste Repository," Space.com, [cited 22 Aug. 2002].

[16]McCurdy, H. E., *Space and the American Imagination*, Smithsonian Inst. Press, Washington, DC, 1997, pp. 235,236.

[17]Rolston, H., "The Preservation of Natural Value in the Solar System," *Beyond Spaceship Earth: Environmental Ethics and the Solar System*, edited by E. C. Hargrove, Sierra Club Books, San Francisco, 1986, p. 171.

[18]Rolston, H., "The Preservation of Natural Value in the Solar System," *Beyond Spaceship Earth: Environmental Ethics and the Solar System*, edited by E. C. Hargrove, Sierra Club Books, San Francisco, 1986, p. 147.

[19]Williamson, M., "Protection of the Space Environment Under the Outer Space Treaty," International Inst. of Space Law, IISL-97-IISL.4.02, Oct. 1997.

[20]Almar, I., "Protection of the Lifeless Environment in the Solar System," Committee on Space Research, COSPAR02-A-00485, Oct. 2002.

[21]Steiner, R. G., "Designating Earth's Moon as a United Nations World Heritage Site—Permanently Protected from Commercial or Military Use," International Academy of Astronautics, IAA-02-IAA.8.1.04, Oct. 2002.

[22]Sterns, P. M., and Tennen, L. I., "Protection of Celestial Environments Through Planetary Quarantine Requirements," *Proceedings of the 23rd Colloquium on the Law of Outer Space*, AIAA, New York, 1980, p. 111.

[23]IAA/IISL Scientific-Legal Round Table on Protection of the Space Environment, 50th IAF Congress, Oct. 1999.

[24]Williamson, M., "Protection of the Space Environment: the First Small Steps," Committee on Space Research, COSPAR02-A-01364/PPP1-0014-02, 2002; *Advances in Space Research*, Vol. 34, 2004, pp. 2338–2343, URL: www.sciencedirect.com.

[25]Williamson, M., "Space Debris: Prevention is Better Than Cure," Editorial, *Space and Communications*, Vol. 12, No. 6, Nov./Dec. 1996, p. 2.

[26]"Orbital Express Refuelling Satellite," *Space News*, 17 Feb. 2003, p. 20.

[27]Berger, B., "Langley Designs Roadmap for a Space Boom Town," *Space News*, Vol. 13, 20 May 2002, p. 8.

[28]Walters, H., *Expedition Venus*, Faber Fanfares edition, Faber and Faber, London, 1962.

[29]Almar, I., and Horvath, A., "Do We Need 'Environmental Protection' in the Solar System?," International Academy of Astronautics, Paper IAA-89-634, Oct. 1989.

[30]Almar, I., "Protection of the Planetary Environment—The Point of View of an Astronomer," International Academy of Astronautics, Paper IAA-99-IAA.7.1.02, Oct. 1999.

[31]Almar, I., "What Could COSPAR Do to Protect the Planetary and Space Environment?," *Advances in Space Research*, Vol. 30, No. 6, 2002, pp. 1577–1581.

[32]Reijnen, B. C. M., "The Pollution of Outer Space," *Environmental Law Review*, Vol. 4/5, Oct. 1993, pp. 117–121.

[33]Perek, L., "Planetary Protection: Lessons Learned," Committee on Space Research, COSPAR02-A-00371, Oct. 2002.

[34]Perek, L., "Space Traffic Management Makes Sense," IAF IAC-04-IAA.5.12.4.02, Oct. 2004.

[35]Reijnen, B. C. M., "The Far Side of the Moon and the Case for an Outer Space Authority," *International Space Review*, Dec. 2004, pp. 4,5, URL: www.satellite-evolution.com.

[36]Williamson, R. A., "The Debate Over Space Control," *International Space Review*, Jan. 2005, p. 5, URL: www.satellite-evolution.com.

[37]Carson, R., *Silent Spring*, Houghton Mifflin, Boston, 1962.

[38]Shackleton, L., "Introduction," *Silent Spring*, R. Carson, Penguin Books, Harmondsworth, England, UK, 1965, p. 16.

Appendix A

TEXT OF COSPAR PLANETARY PROTECTION POLICY (http://www.cosparhq.org/scistr/PPPolicy.htm)

COSPAR PLANETARY PROTECTION POLICY

(20 October 2002; Amended 24 March 2005)

APPROVED BY THE BUREAU AND COUNCIL, WORLD SPACE COUNCIL, HOUSTON, TEXAS, USA

(Prepared by the COSPAR/IAU Workshop on Planetary Protection, 4/02 with updates 10/02)

PREAMBLE

Noting that COSPAR has concerned itself with questions of biological contamination and spaceflight since its very inception, and

noting that Article IX of the Treaty on Principles Governing the Activities of States in the Exploration and Use of Outer Space, Including the Moon and Other Celestial Bodies (also known as the UN Space Treaty of 1967) states that:

> States Parties to the Treaty shall pursue studies of outer space, including the Moon and other celestial bodies, and conduct exploration of them so as to avoid their harmful contamination and also adverse changes in the environment of the Earth resulting from the introduction of extraterrestrial matter, and where necessary, shall adopt appropriate measures for this purpose (UN 1967).

therefore, COSPAR maintains and promulgates this planetary protection policy for the reference of spacefaring nations, both as an international standard on procedures to avoid organic-constituent and biological contamination in space exploration, and to provide accepted guidelines in this area to guide compliance with the wording of this UN Space Treaty and other relevant international agreements.

POLICY

COSPAR,

Referring to COSPAR Resolutions 26.5 and 26.7 of 1964, the Report of the Consultative Group on Potentially Harmful Effects of Space Experiments of 1966, the Report of the same Group of 1967, and the Report of the COSPAR/IAU Workshop of 2002,

notes with appreciation and interest the extensive work done by the Panel on Standards for Space Probe Sterilization and its successors the Panel on Planetary Quarantine and the Panel on Planetary Protection and

accepts that for certain space mission/target body combinations, controls on contamination shall be imposed in accordance with a specified range of requirements, based on the following policy statement:

> Although the existence of life elsewhere in the solar system may be unlikely, the conduct of scientific investigations of possible extra-terrestrial life forms, precursors, and remnants must not be jeopardized. In addition, the Earth must be protected from the potential hazard posed by extraterrestrial matter carried by a spacecraft returning from another planet. Therefore, for certain space mission/target planet com-binations, controls on contamination shall be imposed, in accordance with issuances implementing this policy (DeVincenzi et al. 1983).

The five categories for target body/mission type combinations and their respective suggested ranges of requirements are described as follows, and in Table 1. Assignment of categories for specific mission/body combinations is to be determined by the best multidisciplinary scientific advice. For new determinations not covered by this policy, such advice should be obtained through the auspices of the Member National Scientific Institutions of COSPAR. In case such advice is not available, COSPAR will consider providing such advice through an *ad hoc* multidisciplinary committee formed in consultation with its Member National Scientific Institutions and International Scientific Unions:

Category I includes any mission to a target body which is not of direct interest for understanding the process of chemical evolution or the origin of life. No protection of such bodies is warranted and no planetary protection requirements are imposed by this policy.

Category II missions comprise all types of missions to those target bodies where there is significant interest relative to the process of chemical evolution and the origin of life, but where there is only a remote chance that contamination carried by a spacecraft could jeopardize future explora-tion. The requirements are for simple documentation only. Preparation of a short planetary protection plan is required for these flight projects primarily to outline intended or potential impact targets, brief Pre- and Post-launch analyses detailing impact strategies, and a Post-encounter and End-of-Mission Report which will provide the location of impact if such an event occurs. Solar system bodies considered to be classified as Category II are listed in the Appendix to this document.

Category III missions comprise certain types of missions (mostly flyby and orbiter) to a target body of chemical evolution and/or origin of life interest or for which scientific opinion provides a significant chance of contamination which could jeopardize a future biological experiment. Requirements will consist of documentation (more involved than Category II) and some implementing procedures, including trajectory biasing, the use of cleanrooms during spacecraft assembly and testing, and possibly bio-burden reduction. Although no impact is intended for Category III missions, an inventory of bulk constituent organics is required if the probability of impact is significant. Category III specifications for selected solar system bodies are set forth in the Appendix to this document. Solar system bodies considered to be classified as Category III also are listed in the Appendix.

Category IV missions comprise certain types of missions (mostly probe and lander) to a target body of chemical evolution and/or origin of life interest or for which scientific opinion provides a significant chance of contamination which could jeopardize future biological experiments. Requirements imposed include rather detailed documentation (more involved than Category III), including a bioassay to enumerate the bio-burden, a probability of contamination analysis, an inventory of the bulk constituent organics and an increased number of implementing procedures. The implementing procedures required may include trajectory biasing, cleanrooms, bioload reduction, possible partial sterilization of the direct contact hardware and a bioshield for that hardware. Generally, the requirements and compliance are similar to *Viking*, with the exception of complete lander/probe sterilization. Category IV specifications for selected solar system bodies are set forth in the Appendix to this document. Solar system bodies considered to be classified as Category IV also are listed in the Appendix.

Category V missions comprise all Earth-return missions. The concern for these missions is the protection of the terrestrial system, the Earth and the Moon. (The Moon must be protected from back contamination to retain freedom from planetary protection requirements on Earth–Moon travel.) For solar system bodies deemed by scientific opinion to have no indigenous life forms, a subcategory "unrestricted Earth return" is defined. Missions in this subcategory have planetary protection requirements on the outbound phase only, corresponding to the category of that phase (typically Category I or II). For all other Category V missions, in a subcategory defined as "restricted Earth return," the highest degree of concern is expressed by the absolute prohibition of destructive impact upon return, the need for

Table A.1   Proposed categories for solar-system bodies and types of missions[2-6]

	Category I	Category II	Category III	Category IV	Category V
Type of mission	Any but Earth return	Any but Earth return	No direct contact (flyby, some orbiters)	Direct contact (lander, probe, some orbiters)	Earth return
Target body Degree of concern	See appendix None	See appendix Record of planned impact probability and contamination control measures	See appendix Limit on impact probability Passive bioload control	See appendix Limit on probability of nonnominal impact Limit on bioload (active control)	See appendix If *restricted* Earth return: 1) No impact on Earth or Moon; 2) Returned hardware sterile; 3) Containment of any sample
Representative range of requirements	None	Documentation only (all brief): 1) PP plan 2) Prelaunch report 3) Postlaunch report 4) Post-encounter report	Documentation (Category II plus) 1) Contamination control 2) Organics inventory (as necessary)	Documentation (Category II plus) 1) P_c analysis plan 2) Microbial reduction plan 3) Microbial assay plan	*Outbound* Same category as target body/outbound mission

5) End-of-mission report

Implementing procedures such as:
1) Trajectory biasing
2) Cleanroom
3) Bioload reduction (as necessary)

4) Organics inventory

Implementing procedures such as:
1) Trajectory biasing
2) Cleanroom
3) Bioload reduction
4) Partial sterilization of contacting hardware (as necessary)
5) Bioshield monitoring of bioload via bioassay

Inbound

If *restricted* Earth return:
1) Documentation (Category II plus)
2) P_c analysis plan
3) Microbial reduction plan
4) Microbial assay plan
5) Trajectory biasing
6) Sterile or contained returned hardware
7) Continual monitoring of project activities
8) Project advanced studies/research.

If unrestricted Earth return: None

containment throughout the return phase of all returned hardware which directly contacted the target body or unsterilized material from the body, and the need for containment of any unsterilized sample collected and returned to Earth. Post-mission, there is a need to conduct timely analyses of the unsterilized sample collected and returned to Earth, under strict containment, and using the most sensitive techniques. If any sign of the existence of a nonterrestrial replicating entity is found, the returned sample must remain contained unless treated by an effective sterilizing procedure. Category V concerns are reflected in requirements that encompass those of Category IV plus a continuing monitoring of project activities, studies and research (i.e., in sterilization procedures and containment techniques).

FURTHER, COSPAR

Recommends that COSPAR members provide information to COSPAR within a reasonable time not to exceed six months after launch about the procedures and computations used for planetary protection for each flight and again within one year after the end of a solar-system exploration mission about the areas of the target(s) which may have been subject to contamination. COSPAR will maintain a repository of these reports, make them available to the public, and annually deliver a record of these reports to the Secretary General of the United Nations. For multinational missions, it is suggested that the lead partner should take the lead in submitting these reports.

Reports should include, but not be limited to, the following information:
1. The estimated biological burden at launch, the methods used to obtain the estimate (e.g., assay techniques applied to spacecraft or a proxy), and the statistical uncertainty in the estimate.
2. The probable composition (identification) of the biological burden for Category IV missions, and for Category V "restricted Earth return" missions.
3. Methods used to control the biological burden, decontaminate and/or sterilize the space flight hardware.
4. The organic inventory of all impacting or landed spacecraft or space-craft-components, for quantities exceeding 1 kg.
5. Intended minimum distance from the surface of the target body for launched components, for those vehicles not intended to land on the body.
6. Approximate orbital parameters, expected or realized, for any vehicle which is intended to be placed in orbit around a solar system body.
7. For the end-of-mission, the disposition of the spacecraft and all of its major components, either in space or for landed components by position (or estimated position) on a planetary surface.

(COSPAR 1969, 1984, 1994; Rummel et al. 2002)

REFERENCES

COSPAR. COSPAR RESOLUTION 26.5, *COSPAR Information Bulletin* **20**, 25–26, 1964.

COSPAR. COSPAR DECISION No. 16, *COSPAR Information Bulletin* **50**,15–16, 1969.

COSPAR. COSPAR DECISION No. 9/76, *COSPAR Information Bulletin* **76**, 14, 1976.

COSPAR. COSPAR INTERNAL DECISION No. 7/84, Promulgated by COSPAR Letter 84/692-5.12.-G. 18 July 1984, 1984.

COSPAR. COSPAR DECISION No. 1/94, *COSPAR Information Bulletin* **131**, 30, 1994.

DeVincenzi, D. L., P. D. Stabekis, and J. B. Barengoltz, A proposed new policy for planetary protection, *Adv. Space Res.* **3**, #8, 13, 1983.

DeVincenzi, D. L., P. D. Stabekis, and J. Barengoltz, Refinement of planetary protection policy for Mars missions, *Adv. Space Res.* **18**, #1/2, 314, 1994.

Rummel, J. D., et al. Report of the COSPAR/IAU Workshop on Planetary Protection, COSPAR, Paris, France, 2002.

Space Studies Board, National Research Council (US), *Evaluating the Biological Potential in Samples Returned from Planetary Satellites and Small Solar System Bodies*, Task Group on Sample Return From Small Solar System Bodies, National Academy of Sciences, Washington, D.C., 1998.

Space Studies Board, National Research Council (US), *Preventing the Forward Contamination of Europa*, Task Group on the Forward Contamination of Europa, National Academy of Sciences, Washington, D.C., 2000.

United Nations, Treaty on principles governing the activities of states in the exploration and use of outer space, including the moon and other celestial bodies, Article IX, U.N. Doc. A/RES/2222/(XXI) 25 Jan 1967; TIAS No. 6347, 1967.

IMPLEMENTATION GUIDELINES AND CATEGORY SPECIFICATIONS FOR INDIVIDUAL TARGET BODIES (VERSION 24, MARCH 2005)

IMPLEMENTATION GUIDELINES ON THE USE OF CLEAN-ROOM TECHNOLOGY FOR OUTER-PLANET MISSIONS

COSPAR,

Noting that in the exploration of the outer planets, the probabilities of growth of contaminating terrestrial micro-organisms are extremely low, reflecting the fact that the environments of these planets appear hostile to all known biological processes,

noting also that these environments do not preclude the possibility of *indigenous* life forms in some of these environments,

recognizing that the search for life is a potentially valid objective in the exploration of the outer solar system,

recognizing that the organic chemistry of these bodies remains of paramount importance to our understanding of the process of chemical evolution and its relationship to the origin of life,

recognizing that study of the processes of the pre-biotic organic syntheses under natural conditions must not be jeopardized,

recommends the use of the best available clean-room technology, comparable with that employed for the *Viking* mission, for all missions to the outer planets and their satellites (COSPAR 1976).

NUMERICAL IMPLEMENTATION GUIDELINES FOR FORWARD CONTAMINATION CALCULATIONS

To the degree that numerical guidelines are required to support the overall policy objectives of this document, and except where numerical requirements are otherwise specified, the guideline to be used is that the probability that a planetary body will be contaminated during the period of exploration should be no more than 1×10^{-3}. The period of exploration can be assumed to be no less than 50 years after a Category III or IV mission arrives at its protected target. No specific format for probability of contamination calculations is specified.

IMPLEMENTATION GUIDELINES FOR CATEGORY V MISSIONS

If during the course of a Category V mission there is a change in the circumstances that led to its classification, or a mission failure, e.g.:

- New data or scientific opinion arise that would lead to the reclassification of a mission classified as "Unrestricted Earth return" to "Restricted Earth return," and safe return of the sample cannot be assured, OR
- The sample containment system of a mission classified as "Restricted Earth return" is thought to be compromized, and sample sterilization is impossible,

then the sample to be returned shall be abandoned, and if already collected the spacecraft carrying the sample must not be allowed to return to the Earth or the Moon.

Category-Specific Listing of Target Body/Mission Types

Category I: Flyby, Orbiter, Lander: Venus; Moon; Undifferentiated, metamorphosed asteroids; others TBD

Category II: Flyby, Orbiter, Lander: Comets; Carbonaceous Chondrite Asteroids; Jupiter; Saturn; Uranus; Neptune; Pluto/Charon; Kuiper-Belt Objects; others TBD
Category III: Flyby, Orbiters: Mars; Europa; others TBD
Category IV: Lander Missions: Mars; Europa; others TBD
Category V: Any Earth-return mission. "Restricted Earth return": Mars; Europa; others TBD; "Unrestricted Earth return": Moon; others TBD.

CATEGORY III/IV/V REQUIREMENTS FOR MARS

Missions to Mars

Category III. Mars orbiters will not be required to meet orbital lifetime requirements* if they achieve bioburden levels equivalent to the *Viking* lander pre-sterilization total bioburden. (*Defined as 20 years after launch at greater than or equal to 99% probability, and 50 years after launch at greater than or equal to 95% probability.) (DeVincenzi et al. 1994.)

Category IV for Mars is subdivided into IVa, IVb, and IVc:

Category IVa. Lander systems not carrying instruments for the investigations of extant martian life are restricted to a biological burden no greater than *Viking* lander pre-sterilization levels

Category IVb. For lander systems designed to investigate extant martian life, all of the requirements of Category IVa apply, along with the following requirement:

The entire landed system must be sterilized at least to *Viking* post-sterilization biological burden levels, or to levels of biological burden reduction driven by the nature and sensitivity of the particular life-detection experiments, whichever are more stringent

OR

The subsystems which are involved in the acquisition, delivery, and analysis of samples used for life detection must be sterilized to these levels, and a method of preventing recontamination of the sterilized subsystems and the contamination of the material to be analyzed is in place.

Category IVc. For missions which investigate martian special regions (see definition below), even if they do not include life detection experiments, all of the requirements of Category IVa apply, along with the following requirement:

Case 1. If the landing site is within the special region, the entire landed system shall be sterilized at least to the *Viking* post-sterilization biological burden levels.

Case 2. If the special region is accessed though horizontal or vertical mobility, either the entire landed system shall be sterilized to the *Viking*

post-sterilization biological burden levels, **OR** the subsystems which directly contact the special region shall be sterilized to these levels, and a method of preventing their recontamination prior to accessing the special region shall be provided.

If an off-nominal condition (such as a hard landing) would cause a high probability of inadvertent biological contamination of the special region by the spacecraft, the entire landed system must be sterilized to the *Viking* post-sterilization biological burden levels.

Definition of "Special Region"

A Special Region is defined as a region within which terrestrial organisms are likely to propagate, **OR** a region which is interpreted to have a high potential for the existence of extant martian life forms.

Given current understanding, this is apply to regions where liquid water is present or may occur. Specific examples include but are not limited to:

- Subsurface access in an area and to a depth where the presence of liquid water is probable
- Penetrations into the polar caps
- Areas of hydrothermal activity.

Sample Return Missions from Mars

Category V. The Earth return mission is classified, "Restricted Earth return."

- Unless specifically exempted, the outbound leg of the mission shall meet Category IVb requirements. This provision is intended to avoid "false positive" indications in a life-detection and hazard-determination protocol, or in the search for life in the sample after it is returned. A "false positive" could prevent distribution of the sample from containment and could lead to unnecessary increased rigor in the requirements for all later Mars missions.
- Unless the sample to be returned is subjected to an accepted and approved sterilization process, the sample container must be sealed after sample acquisition, and a redundant, fail-safe containment with a method for verification of its operation before Earth-return shall be required. For unsterilized samples, the integrity of the flight containment system shall be maintained until the sample is transferred to containment in an appropriate receiving facility.
- The mission and the spacecraft design must provide a method to "break the chain of contact" with Mars. No uncontained hardware that contacted Mars, directly or indirectly, shall be returned to Earth. Isolation of such hardware from the Mars environment shall be provided during sample container loading into the containment system, launch from Mars, and any in-flight transfer operations required by the mission.

- Reviews and approval of the continuation of the flight mission shall be required at three stages: 1) prior to launch from Earth; 2) prior to leaving Mars for return to Earth; and 3) prior to commitment to Earth re-entry.
- For unsterilized samples returned to Earth, a program of life detection and biohazard testing, or a proven sterilization process, shall be undertaken as an absolute precondition for the controlled distribution of any portion of the sample.

CATEGORY III/IV/V REQUIREMENTS FOR EUROPA

Missions to Europa

Category III and IV. Requirements for Europa flybys, orbiters and landers, including bioburden reduction, shall be applied in order to reduce the probability of inadvertent contamination of an europan ocean to less than 1×10^{-4} per mission. These requirements will be refined in future years, but the calculation of this probability should include a conservative estimate of poorly known parameters, and address the following factors, at a minimum:

- Bioburden at launch
- Cruise survival for contaminating organisms
- Organism survival in the radiation environment adjacent to Europa
- Probability of landing on Europa
- The mechanisms and timescales of transport to the europan subsurface
- Organism survival and proliferation before, during, and after subsurface transfer

Preliminary calculations of the probability of contamination suggest that bioburden reduction will likely be necessary even for Europa orbiters (Category III) as well as for landers, requiring the use of cleanroom technology and the cleanliness of all parts before assembly, and the monitoring of spacecraft assembly facilities to understand the bioload and its microbial diversity, including specific problematic species. Specific methods should be developed to eradicate problematic species. Methods of bioburden reduction should reflect the type of environments found on Europa, focusing on Earth extremophiles most likely to survive on Europa, such as cold and radiation tolerant organisms (SSB 2000).

Sample Return Missions from Europa

Category V. The Earth return mission is classified, "Restricted Earth return."

- Unless specifically exempted, the outbound leg of the mission shall meet the contamination control requirements given above. This provision should avoid "false positive" indications in a life-detection and hazard-determination protocol, or in the search for life in the sample after it is returned. A "false positive" could prevent distribution of the sample from containment and could lead to unnecessary increased rigor in the requirements for all later Europa missions.

- Unless the sample to be returned is subjected to an accepted and approved sterilization process, the sample container must be sealed after sample acquisition, and a redundant, fail-safe containment with a method for verification of its operation before Earth-return shall be required. For unsterilized samples, the integrity of the flight containment system shall be maintained until the sample is transferred to containment in an appropriate receiving facility.
- The mission and the spacecraft design must provide a method to "break the chain of contact" with Europa. No uncontained hardware that contacted Europa, directly or indirectly, shall be returned to Earth. Isolation of such hardware from the europan environment shall be provided during sample container loading into the containment system, launch from Europa, and any in-flight transfer operations required by the mission.
- Reviews and approval of the continuation of the flight mission shall be required at three stages: 1) prior to launch from Earth; 2) prior to leaving Europa for return to Earth; and 3) prior to commitment to Earth re-entry.
- For unsterilized samples returned to Earth, a program of life detection and biohazard testing, or a proven sterilization process, shall be undertaken as an absolute precondition for the controlled distribution of any portion of the sample (SSB 1998).

CATEGORY REQUIREMENTS FOR SMALL SOLAR SYSTEM BODIES

Missions to Small Solar System Bodies

Category I, II, III, or IV. The small bodies of the solar system not elsewhere discussed in this policy represent a very large class of objects. Imposing forward contamination controls on these missions is not warranted except on a case-by-case basis, so most such missions should reflect Categories I or II. Further elaboration of this requirement is anticipated.

Sample Return Missions from Small Solar System Bodies

Category V. Determination as to whether a mission is classified "Restricted Earth return" or not shall be undertaken with respect to the best multidisciplinary scientific advice, using the framework presented in the 1998 report of the US National Research Council's Space Studies Board entitled, *Evaluating the Biological Potential in Samples Returned from Planetary Satellites and Small Solar System Bodies: Framework for Decision Making* (SSB 1998). Specifically, such a determination shall address the following six questions for each body intended to be sampled:

1. Does the preponderance of scientific evidence indicate that there was never liquid water in or on the target body?

2. Does the preponderance of scientific evidence indicate that metabolically useful energy sources were never present?

3. Does the preponderance of scientific evidence indicate that there was never sufficient organic matter (or CO_2 or carbonates and an appropriate source of reducing equivalents) in or on the target body to support life?

4. Does the preponderance of scientific evidence indicate that subsequent to the disappearance of liquid water, the target body has been subjected to extreme temperatures (i.e., $>160°C$)?

5. Does the preponderance of scientific evidence indicate that there is or was sufficient radiation for biological sterilization of terrestrial life forms?

6. Does the preponderance of scientific evidence indicate that there has been a natural influx to Earth, e.g., via meteorites, of material equivalent to a sample returned from the target body?

For containment procedures to be necessary ("Restricted Earth return"), an answer of "no" or "uncertain" needs to be returned to all six questions.

For missions determined to be Category V, "Restricted Earth return," the following requirements shall be met:

- Unless specifically exempted, the outbound leg of the mission shall meet contamination control requirements to avoid "false positive" indications in a life-detection and hazard-determination protocol, or in any search for life in the sample after it is returned. A "false positive" could prevent distribution of the sample from containment and could lead to unnecessary increased rigor in the requirements for all later missions to that body.

- Unless the sample to be returned is subjected to an accepted and approved sterilization process, the sample container must be sealed after sample acquisition, and a redundant, fail-safe containment with a method for verification of its operation before Earth-return shall be required. For unsterilized samples, the integrity of the flight containment system shall be maintained until the sample is transferred to containment in an appropriate receiving facility.

- The mission and the spacecraft design must provide a method to "break the chain of contact" with the small body. No uncontained hardware that contacted the body, directly or indirectly, shall be returned to Earth. Isolation of such hardware from the body's environment shall be provided during sample container loading into the containment system, launch from the body, and any in-flight transfer operations required by the mission.

- Reviews and approval of the continuation of the flight mission shall be required at three stages: 1) prior to launch from Earth; 2) prior to leaving the body or its environment for return to Earth; and 3) prior to commitment to Earth re-entry.

- For unsterilized samples returned to Earth, a program of life detection and biohazard testing, or a proven sterilization process, shall be undertaken as an absolute precondition for the controlled distribution of any portion of the sample (SSB 1998).

Texts of Selected Space Law Treaties—The Outer Space Treaty of 1967 and the Moon Agreement of 1984 (http://www.oosa.unvienna.org/SpaceLaw/ treaties.html)

Treaty on Principles Governing the Activities of States in the Exploration and Use of Outer Space, Including the Moon and Other Celestial Bodies

The States Parties to this Treaty,

Inspired by the great prospects opening up before mankind as a result of man's entry into outer space,

Recognizing the common interest of all mankind in the progress of the exploration and use of outer space for peaceful purposes,

Believing that the exploration and use of outer space should be carried on for the benefit of all peoples irrespective of the degree of their economic or scientific development,

Desiring to contribute to broad international co-operation in the scientific as well as the legal aspects of the exploration and use of outer space for peaceful purposes,

Believing that such co-operation will contribute to the development of mutual understanding and to the strengthening of friendly relations between States and peoples,

Recalling resolution 1962 (XVIII), entitled "Declaration of Legal Principles Governing the Activities of States in the Exploration and Use of Outer Space," which was adopted unanimously by the United Nations General Assembly on 13 December 1963,

Recalling resolution 1884 (XVIII), calling upon States to refrain from placing in orbit around the earth any objects carrying nuclear weapons or any other kinds of weapons of mass destruction or from installing such weapons on celestial bodies, which was adopted unanimously by the United Nations General Assembly on 17 October 1963,

Taking account of United Nations General Assembly resolution 110 (II) of 3 November 1947, which condemned propaganda designed or likely to provoke or encourage any threat to the peace, breach of the peace or act of aggression, and considering that the aforementioned resolution is applicable to outer space,

Convinced that a Treaty on Principles Governing the Activities of States in the Exploration and Use of Outer Space, including the Moon and Other Celestial Bodies, will further the purposes and principles of the Charter of the United Nations,

Have agreed on the following:

ARTICLE I

The exploration and use of outer space, including the Moon and other celestial bodies, shall be carried out for the benefit and in the interests of all countries, irrespective of their degree of economic or scientific development, and shall be the province of all mankind.

Outer space, including the Moon and other celestial bodies, shall be free for exploration and use by all States without discrimination of any kind, on a basis of equality and in accordance with international law, and there shall be free access to all areas of celestial bodies.

There shall be freedom of scientific investigation in outer space, including the moon and other celestial bodies, and States shall facilitate and encourage international co-operation in such investigation.

ARTICLE II

Outer space, including the Moon and other celestial bodies, is not subject to national appropriation by claim of sovereignty, by means of use or occupation, or by any other means.

ARTICLE III

States Parties to the Treaty shall carry on activities in the exploration and use of outer space, including the moon and other celestial bodies, in accordance with international law, including the Charter of the United Nations, in the interest of maintaining international peace and security and promoting international co-operation and understanding.

ARTICLE IV

States Parties to the Treaty undertake not to place in orbit around the earth any objects carrying nuclear weapons or any other kinds of weapons of mass destruction, install such weapons on celestial bodies, or station such weapons in outer space in any other manner.

The Moon and other celestial bodies shall be used by all States Parties to the Treaty exclusively for peaceful purposes. The establishment of military bases, installations and fortifications, the testing of any type of weapons and the conduct of military manoeuvres on celestial bodies shall be forbidden. The use of military personnel for scientific research or for any other peaceful purposes shall not be prohibited. The use of any equipment or facility necessary for peaceful exploration of the Moon and other celestial bodies shall also not be prohibited.

ARTICLE V

States Parties to the Treaty shall regard astronauts as envoys of mankind in outer space and shall render to them all possible assistance in the event of accident, distress, or emergency landing on the territory of another State Party or on the high seas. When astronauts make such a landing, they shall be safely and promptly returned to the State of registry of their space vehicle.

In carrying on activities in outer space and on celestial bodies, the astronauts of one State Party shall render all possible assistance to the astronauts of other States Parties.

States Parties to the Treaty shall immediately inform the other States Parties to the Treaty or the Secretary-General of the United Nations of any phenomena they discover in outer space, including the Moon and other celestial bodies, which could constitute a danger to the life or health of astronauts.

ARTICLE VI

States Parties to the Treaty shall bear international responsibility for national activities in outer space, including the Moon and other celestial bodies, whether such activities are carried on by governmental agencies or by non-governmental entities, and for assuring that national activities are carried out in conformity with the provisions set forth in the present Treaty. The activities of non-governmental entities in outer space, including the Moon and other celestial bodies, shall require authorization and continuing supervision by the appropriate State Party to the Treaty. When activities are carried on in outer space, including the Moon and other celestial bodies, by an international organization, responsibility for compliance with this Treaty shall be borne both by the international organization and by the States Parties to the Treaty participating in such organization.

ARTICLE VII

Each State Party to the Treaty that launches or procures the launching of an object into outer space, including the Moon and other celestial bodies, and each State Party from whose territory or facility an object is launched, is internationally liable for damage to another State Party to the Treaty or to its natural or juridical persons by such object or its component parts on the Earth, in air or in outer space, including the Moon and other celestial bodies.

ARTICLE VIII

A State Party to the Treaty on whose registry an object launched into outer space is carried shall retain jurisdiction and control over such object, and over any personnel thereof, while in outer space or on a celestial body. Ownership of objects launched into outer space, including objects landed or constructed on a celestial body, and of their component parts, is not affected by their presence in outer space or on a celestial body or by their return to the Earth. Such objects or component parts found beyond the limits of the State Party to the Treaty on whose registry they are carried shall be returned to that State Party, which shall, upon request, furnish identifying data prior to their return.

ARTICLE IX

In the exploration and use of outer space, including the Moon and other celestial bodies, States Parties to the Treaty shall be guided by the principle of co-operation and mutual assistance and shall conduct all their activities in outer space, including the Moon and other celestial bodies, with due regard to the corresponding interests of all other States Parties to the Treaty. States Parties to the Treaty shall pursue studies of outer space, including the Moon and other celestial bodies, and conduct exploration of them so as to avoid their harmful contamination and also adverse changes in the environment of the Earth resulting from the introduction of extraterrestrial matter and, where necessary, shall adopt appropriate measures for this purpose. If a State Party to the Treaty has reason to believe that an activity or experiment planned by it or its nationals in outer space, including the Moon and other celestial bodies, would cause potentially harmful interference with activities of other States Parties in the peaceful exploration and use of outer space, including the Moon and other celestial bodies, it shall undertake appropriate international consultations before proceeding with any such activity or experiment. A State Party to the Treaty which has reason to believe that an activity or experiment planned by another State Party in outer space, including the Moon and other celestial bodies, would cause potentially harmful interference with activities in the peaceful exploration and use of outer space, including the Moon and other celestial bodies, may request consultation concerning the activity or experiment.

ARTICLE X

In order to promote international co-operation in the exploration and use of outer space, including the Moon and other celestial bodies, in conformity with the purposes of this Treaty, the States Parties to the Treaty shall consider on a basis of equality any requests by other States Parties to the Treaty to be afforded an opportunity to observe the flight of space objects launched by those States. The nature of such an opportunity for observation and the conditions under which it could be afforded shall be determined by agreement between the States concerned.

ARTICLE XI

In order to promote international co-operation in the peaceful exploration and use of outer space, States Parties to the Treaty conducting activities in outer space, including the Moon and other celestial bodies, agree to inform the Secretary-General of the United Nations as well as the public and the international scientific community, to the greatest extent feasible and practicable, of the nature, conduct, locations and results of such activities. On receiving the said information, the Secretary-General of the United Nations should be prepared to disseminate it immediately and effectively.

ARTICLE XII

All stations, installations, equipment and space vehicles on the Moon and other celestial bodies shall be open to representatives of other States Parties to the Treaty on a basis of reciprocity. Such representatives shall give reasonable advance notice of a projected visit, in order that appropriate consultations may be held and that maximum precautions may be taken to assure safety and to avoid interference with normal operations in the facility to be visited.

ARTICLE XIII

The provisions of this Treaty shall apply to the activities of States Parties to the Treaty in the exploration and use of outer space, including the Moon and other celestial bodies, whether such activities are carried on by a single State Party to the Treaty or jointly with other States, including cases where they are carried on within the framework of international intergovernmental organizations.

Any practical questions arising in connection with activities carried on by international intergovernmental organizations in the exploration and use of outer space, including the Moon and other celestial bodies, shall be resolved by the States Parties to the Treaty either with the appropriate international organization or with one or more States members of that international organization, which are Parties to this Treaty.

ARTICLE XIV

1. This Treaty shall be open to all States for signature. Any State which does not sign this Treaty before its entry into force in accordance with paragraph 3 of this article may accede to it at anytime.
2. This Treaty shall be subject to ratification by signatory States. Instruments of ratification and instruments of accession shall be deposited with the Governments of the United Kingdom of Great Britain and Northern Ireland, the Union of Soviet Socialist Republics and the United States of America, which are hereby designated the Depositary Governments.
3. This Treaty shall enter into force upon the deposit of instruments of ratification by five Governments including the Governments designated as Depositary Governments under this Treaty.
4. For States whose instruments of ratification or accession are deposited subsequent to the entry into force of this Treaty, it shall enter into force on the date of the deposit of their instruments of ratification or accession.
5. The Depositary Governments shall promptly inform all signatory and acceding States of the date of each signature, the date of deposit of each instrument of ratification of and accession to this Treaty, the date of its entry into force and other notices.
6. This Treaty shall be registered by the Depositary Governments pursuant to Article 102 of the Charter of the United Nations.

ARTICLE XV

Any State Party to the Treaty may propose amendments to this Treaty. Amendments shall enter into force for each State Party to the Treaty accepting the amendments upon their acceptance by a majority of the States Parties to the Treaty and thereafter for each remaining State Party to the Treaty on the date of acceptance by it.

ARTICLE XVI

Any State Party to the Treaty may give notice of its withdrawal from the Treaty one year after its entry into force by written notification to the Depositary Governments. Such withdrawal shall take effect one year from the date of receipt of this notification.

ARTICLE XVII

This Treaty, of which the English, Russian, French, Spanish and Chinese texts are equally authentic, shall be deposited in the archives of the Depositary Governments. Duly certified copies of this Treaty shall be transmitted by the Depositary Governments to the Governments of the signatory and acceding States.

IN WITNESS WHEREOF the undersigned, duly authorized, have signed this Treaty.

DONE in triplicate, at the cities of London, Moscow and Washington, the twenty-seventh day of January, one thousand nine hundred and sixty-seven.

AGREEMENT GOVERNING THE ACTIVITIES OF STATES ON THE MOON AND OTHER CELESTIAL BODIES

The States Parties to this Agreement,

Noting the achievements of States in the exploration and use of the Moon and other celestial bodies,

Recognizing that the Moon, as a natural satellite of the earth, has an important role to play in the exploration of outer space,

Determined to promote on the basis of equality the further development of co-operation among States in the exploration and use of the Moon and other celestial bodies,

Desiring to prevent the Moon from becoming an area of international conflict,

Bearing in mind the benefits which may be derived from the exploitation of the natural resources of the Moon and other celestial bodies,

Recalling the Treaty on Principles Governing the Activities of States in the Exploration and Use of Outer Space, including the Moon and Other Celestial Bodies, the Agreement on the Rescue of Astronauts, the Return of Astronauts and the Return of Objects Launched into Outer Space, the Convention on International Liability for Damage Caused by Space Objects, and the Convention on Registration of Objects Launched into Outer Space,

Taking into account the need to define and develop the provisions of these international instruments in relation to the Moon and other celestial bodies, having regard to further progress in the exploration and use of outer space,

Have agreed on the following:

ARTICLE 1

1. The provisions of this Agreement relating to the Moon shall also apply to other celestial bodies within the solar system, other than the earth, except in so far as specific legal norms enter into force with respect to any of these celestial bodies.
2. For the purposes of this Agreement reference to the Moon shall include orbits around or other trajectories to or around it.

3. This Agreement does not apply to extraterrestrial materials which reach the surface of the Earth by natural means.

ARTICLE 2

All activities on the Moon, including its exploration and use, shall be carried out in accordance with international law, in particular the Charter of the United Nations, and taking into account the Declaration on Principles of International Law concerning Friendly Relations and Co-operation among States in accordance with the Charter of the United Nations, adopted by the General Assembly on 24 October 1970, in the interest of maintaining international peace and security and promoting international co-operation and mutual understanding, and with due regard to the corresponding interests of all other States Parties.

ARTICLE 3

1. The Moon shall be used by all States Parties exclusively for peaceful purposes.
2. Any threat or use of force or any other hostile act or threat of hostile act on the Moon is prohibited. It is likewise prohibited to use the Moon in order to commit any such act or to engage in any such threat in relation to the earth, the Moon, spacecraft, the personnel of spacecraft or man-made space objects.
3. States Parties shall not place in orbit around or other trajectory to or around the Moon objects carrying nuclear weapons or any other kinds of weapons of mass destruction or place or use such weapons on or in the Moon.
4. The establishment of military bases, installations and fortifications, the testing of any type of weapons and the conduct of military manoeuvres on the Moon shall be forbidden. The use of military personnel for scientific research or for any other peaceful purposes shall not be prohibited. The use of any equipment or facility necessary for peaceful exploration and use of the Moon shall also not be prohibited.

ARTICLE 4

1. The exploration and use of the Moon shall be the province of all mankind and shall be carried out for the benefit and in the interests of all countries, irrespective of their degree of economic or scientific development. Due regard shall be paid to the interests of present and future generations as well as to the need to promote higher standards of living and conditions of economic and social progress and development in accordance with the Charter of the United Nations.

2. States Parties shall be guided by the principle of co-operation and mutual assistance in all their activities concerning the exploration and use of the Moon. International co-operation in pursuance of this Agreement should be as wide as possible and may take place on a multilateral basis, on a bilateral basis or through international inter-governmental organizations.

ARTICLE 5

1. States Parties shall inform the Secretary-General of the United Nations as well as the public and the international scientific community, to the greatest extent feasible and practicable, of their activities concerned with the exploration and use of the Moon. Information on the time, purposes, locations, orbital parameters and duration shall be given in respect of each mission to the Moon as soon as possible after launch-ing, while information on the results of each mission, including scien-tific results, shall be furnished upon completion of the mission. In the case of a mission lasting more than sixty days, information on conduct of the mission, including any scientific results, shall be given period-ically, at thirty-day intervals. For missions lasting more than six months, only significant additions to such information need be reported thereafter.

2. If a State Party becomes aware that another State Party plans to operate simultaneously in the same area of or in the same orbit around or trajectory to or around the Moon, it shall promptly inform the other State of the timing of and plans for its own operations.

3. In carrying out activities under this Agreement, States Parties shall promptly inform the Secretary-General, as well as the public and the international scientific community, of any phenomena they discover in outer space, including the Moon, which could endanger human life or health, as well as of any indication of organic life.

ARTICLE 6

1. There shall be freedom of scientific investigation on the Moon by all States Parties without discrimination of any kind, on the basis of equal-ity and in accordance with international law.

2. In carrying out scientific investigations and in furtherance of the pro-visions of this Agreement, the States Parties shall have the right to collect on and remove from the Moon samples of its mineral and other substances. Such samples shall remain at the disposal of those States Parties which caused them to be collected and may be used by them for scientific purposes. States Parties shall have regard to the desirability of making a portion of such samples available to other

interested States Parties and the international scientific community for scientific investigation. States Parties may in the course of scientific investigations also use mineral and other substances of the Moon in quantities appropriate for the support of their missions.

3. States Parties agree on the desirability of exchanging scientific and other personnel on expeditions to or installations on the Moon to the greatest extent feasible and practicable.

ARTICLE 7

1. In exploring and using the Moon, States Parties shall take measures to prevent the disruption of the existing balance of its environment, whether by introducing adverse changes in that environment, by its harmful contamination through the introduction of extra-environmental matter or otherwise. States Parties shall also take measures to avoid harmfully affecting the environment of the Earth through the introduction of extraterrestrial matter or otherwise.

2. States Parties shall inform the Secretary-General of the United Nations of the measures being adopted by them in accordance with paragraph 1 of this article and shall also, to the maximum extent feasible, notify him in advance of all placements by them of radio-active materials on the Moon and of the purposes of such placements.

3. States Parties shall report to other States Parties and to the Secretary-General concerning areas of the Moon having special scientific interest in order that, without prejudice to the rights of other States Parties, consideration may be given to the designation of such areas as international scientific preserves for which special protective arrangements are to be agreed upon in consultation with the competent bodies of the United Nations.

ARTICLE 8

1. States Parties may pursue their activities in the exploration and use of the Moon anywhere on or below its surface, subject to the provisions of this Agreement.

2. For these purposes States Parties may, in particular:
   (a) Land their space objects on the Moon and launch them from the Moon;
   (b) Place their personnel, space vehicles, equipment, facilities, stations and installations anywhere on or below the surface of the Moon.
   Personnel, space vehicles, equipment, facilities, stations and installations may move or be moved freely over or below the surface of the Moon.

3. Activities of States Parties in accordance with paragraphs 1 and 2 of this article shall not interfere with the activities of other States

Parties on the Moon. Where such interference may occur, the States Parties concerned shall undertake consultations in accordance with article 15, paragraphs 2 and 3, of this Agreement.

ARTICLE 9

1. States Parties may establish manned and unmanned stations on the moon. A State Party establishing a station shall use only that area which is required for the needs of the station and shall immediately inform the Secretary-General of the United Nations of the location and purposes of that station. Subsequently, at annual intervals that State shall likewise inform the Secretary-General whether the station continues in use and whether its purposes have changed.
2. Stations shall be installed in such a manner that they do not impede the free access to all areas of the Moon of personnel, vehicles and equipment of other States Parties conducting activities on the Moon in accordance with the provisions of this Agreement or of article I of the Treaty on Principles Governing the Activities of States in the Exploration and Use of Outer Space, including the Moon and Other Celestial Bodies.

ARTICLE 10

1. States Parties shall adopt all practicable measures to safeguard the life and health of persons on the Moon. For this purpose they shall regard any person on the Moon as an astronaut within the meaning of article V of the Treaty on Principles Governing the Activities of States in the Exploration and Use of Outer Space, including the Moon and Other Celestial Bodies and as part of the personnel of a spacecraft within the meaning of the Agreement on the Rescue of Astronauts, the Return of Astronauts and the Return of Objects Launched into Outer Space.
2. States Parties shall offer shelter in their stations, installations, vehicles and other facilities to persons in distress on the Moon.

ARTICLE 11

1. The Moon and its natural resources are the common heritage of mankind, which finds its expression in the provisions of this Agreement, in particular in paragraph 5 of this article.
2. The Moon is not subject to national appropriation by any claim of sovereignty, by means of use or occupation, or by any other means.
3. Neither the surface nor the subsurface of the Moon, nor any part thereof or natural resources in place, shall become property of any State, international intergovernmental or non-governmental

organization, national organization or non-governmental entity or of any natural person. The placement of personnel, space vehicles, equipment, facilities, stations and installations on or below the surface of the Moon, including structures connected with its surface or subsurface, shall not create a right of ownership over the surface or the subsurface of the Moon or any areas thereof. The foregoing provisions are without prejudice to the international regime referred to in paragraph 5 of this article.

4. States Parties have the right to exploration and use of the Moon without discrimination of any kind, on the basis of equality and in accordance with international law and the terms of this Agreement.

5. States Parties to this Agreement hereby undertake to establish an international regime, including appropriate procedures, to govern the exploitation of the natural resources of the Moon as such exploitation is about to become feasible. This provision shall be implemented in accordance with article 18 of this Agreement.

6. In order to facilitate the establishment of the international regime referred to in paragraph 5 of this article, States Parties shall inform the Secretary-General of the United Nations as well as the public and the international scientific community, to the greatest extent feasible and practicable, of any natural resources they may discover on the Moon.

7. The main purposes of the international regime to be established shall include:

    (a) The orderly and safe development of the natural resources of the Moon;
    (b) The rational management of those resources;
    (c) The expansion of opportunities in the use of those resources;
    (d) An equitable sharing by all States Parties in the benefits derived from those resources, whereby the interests and needs of the developing countries, as well as the efforts of those countries which have contributed either directly or indirectly to the exploration of the Moon, shall be given special consideration.

8. All the activities with respect to the natural resources of the Moon shall be carried out in a manner compatible with the purposes specified in paragraph 7 of this article and the provisions of article 6, paragraph 2, of this Agreement.

ARTICLE 12

1. States Parties shall retain jurisdiction and control over their personnel, vehicles, equipment, facilities, stations and installations on the Moon. The ownership of space vehicles, equipment, facilities, stations and installations shall not be affected by their presence on the Moon.

2. Vehicles, installations and equipment or their component parts found in places other than their intended location shall be dealt with in accordance with article 5 of the Agreement on the Rescue of Astronauts, the Return of Astronauts and the Return of Objects Launched into Outer Space.

3. In the event of an emergency involving a threat to human life, States Parties may use the equipment, vehicles, installations, facilities or supplies of other States Parties on the Moon. Prompt notification of such use shall be made to the Secretary-General of the United Nations or the State Party concerned.

ARTICLE 13

A State Party which learns of the crash landing, forced landing or other unintended landing on the Moon of a space object, or its component parts, that were not launched by it, shall promptly inform the launching State Party and the Secretary-General of the United Nations.

ARTICLE 14

1. States Parties to this Agreement shall bear international responsibility for national activities on the Moon, whether such activities are carried on by governmental agencies or by non-governmental entities, and for assuring that national activities are carried out in conformity with the provisions set forth in this Agreement. States Parties shall ensure that non-governmental entities under their jurisdiction shall engage in activities on the Moon only under the authority and continuing supervision of the appropriate State Party.

2. States Parties recognize that detailed arrangements concerning liability for damage caused on the Moon, in addition to the provisions of the Treaty on Principles Governing the Activities of States in the Exploration and Use of Outer Space, including the Moon and Other Celestial Bodies and the Convention on International Liability for Damage Caused by Space Objects, may become necessary as a result of more extensive activities on the Moon. Any such arrangements shall be elaborated in accordance with the procedure provided for in article 18 of this Agreement.

ARTICLE 15

1. Each State Party may assure itself that the activities of other States Parties in the exploration and use of the Moon are compatible with the provisions of this Agreement. To this end, all space vehicles, equipment, facilities, stations and installations on the Moon shall be open to other States Parties. Such States Parties shall give reasonable advance

notice of a projected visit, in order that appropriate consultations may be held and that maximum precautions may be taken to assure safety and to avoid interference with normal operations in the facility to be visited. In pursuance of this article, any State Party may act on its own behalf or with the full or partial assistance of any other State Party or through appropriate international procedures within the framework of the United Nations and in accordance with the Charter.

2. A State Party which has reason to believe that another State Party is not fulfilling the obligations incumbent upon it pursuant to this Agreement or that another State Party is interfering with the rights which the former State has under this Agreement may request consultations with that State Party. A State Party receiving such a request shall enter into such consultations without delay. Any other State Party which requests to do so shall be entitled to take part in the consultations. Each State Party participating in such consultations shall seek a mutually acceptable resolution of any controversy and shall bear in mind the rights and interests of all States Parties. The Secretary-General of the United Nations shall be informed of the results of the consultations and shall transmit the information received to all States Parties concerned.

3. If the consultations do not lead to a mutually acceptable settlement which has due regard for the rights and interests of all States Parties, the parties concerned shall take all measures to settle the dispute by other peaceful means of their choice appropriate to the circumstances and the nature of the dispute. If difficulties arise in connection with the opening of consultations or if consultations do not lead to a mutually acceptable settlement, any State Party may seek the assistance of the Secretary-General, without seeking the consent of any other State Party concerned, in order to resolve the controversy. A State Party which does not maintain diplomatic relations with another State Party concerned shall participate in such consultations, at its choice, either itself or through another State Party or the Secretary-General as intermediary.

ARTICLE 16

With the exception of articles 17 to 21, references in this Agreement to States shall be deemed to apply to any international intergovernmental organization which conducts space activities if the organization declares its acceptance of the rights and obligations provided for in this Agreement and if a majority of the States members of the organization are States Parties to this Agreement and to the Treaty on Principles Governing the Activities of States in the Exploration and Use of Outer Space, including the Moon and Other Celestial Bodies. States members of any such organization which are States Parties to this Agreement shall take all appropriate steps to ensure that the organization makes a declaration in accordance with the foregoing.

ARTICLE 17

Any State Party to this Agreement may propose amendments to the Agreement. Amendments shall enter into force for each State Party to the Agreement accepting the amendments upon their acceptance by a majority of the States Parties to the Agreement and thereafter for each remaining State Party to the Agreement on the date of acceptance by it.

ARTICLE 18

Ten years after the entry into force of this Agreement, the question of the review of the Agreement shall be included in the provisional agenda of the General Assembly of the United Nations in order to consider, in the light of past application of the Agreement, whether it requires revision. However, at any time after the Agreement has been in force for five years, the Secretary-General of the United Nations, as depository, shall, at the request of one third of the States Parties to the Agreement and with the concurrence of the majority of the States Parties, convene a conference of the States Parties to review this Agreement. A review conference shall also consider the question of the implementation of the provisions of article 11, paragraph 5, on the basis of the principle referred to in paragraph 1 of that article and taking into account in particular any relevant technological developments.

ARTICLE 19

1. This Agreement shall be open for signature by all States at United Nations Headquarters in New York.
2. This Agreement shall be subject to ratification by signatory States. Any State which does not sign this Agreement before its entry into force in accordance with paragraph 3 of this article may accede to it at any time. Instruments of ratification or accession shall be deposited with the Secretary-General of the United Nations.
3. This Agreement shall enter into force on the thirtieth day following the date of deposit of the fifth instrument of ratification.
4. For each State depositing its instrument of ratification or accession after the entry into force of this Agreement, it shall enter into force on the thirtieth day following the date of deposit of any such instrument.

5. The Secretary-General shall promptly inform all signatory and acceding States of the date of each signature, the date of deposit of each instrument of ratification or accession to this Agreement, the date of its entry into force and other notices.

Article 20

Any State Party to this Agreement may give notice of its withdrawal from the Agreement one year after its entry into force by written notification to the Secretary-General of the United Nations. Such withdrawal shall take effect one year from the date of receipt of this notification.

Article 21

The original of this Agreement, of which the Arabic, Chinese, English, French, Russian and Spanish texts are equally authentic, shall be deposited with the Secretary-General of the United Nations, who shall send certified copies thereof to all signatory and acceding States.

IN WITNESS WHEREOF the undersigned, being duly authorized thereto by their respective Governments, have signed this Agreement, opened for signature at New York on 18 December 1979.

INDEX

Note: Page numbers in **bold** refer to illustrations; entries in *italics* are titles of publications.

SUPPORTING MATERIALS

A complete list of AIAA publications is available at http://www.aiaa.org.